Springer Proceedings in Mathematics & Statistics

Volume 411

This book series features volumes composed of selected contributions from workshops and conferences in all areas of current research in mathematics and statistics, including data science, operations research and optimization. In addition to an overall evaluation of the interest, scientific quality, and timeliness of each proposal at the hands of the publisher, individual contributions are all refereed to the high quality standards of leading journals in the field. Thus, this series provides the research community with well-edited, authoritative reports on developments in the most exciting areas of mathematical and statistical research today.

João Paulo Almeida · Carla Soares Geraldes ·
Isabel Cristina Lopes · Samuel Moniz ·
José Fernando Oliveira · Alberto Adrego Pinto
Editors

Operational Research

IO 2021—Analytics for a Better World. XXI Congress of APDIO, Figueira da Foz, Portugal, November 7–8, 2021

Springer

Editors
João Paulo Almeida
CeDRI-IPB; Department of Mathematics
School of Technology and Management
Polytechnic Institute of Bragança
Bragança, Portugal

Isabel Cristina Lopes
CEOS.PP; Instituto Superior de Contabilidade e Administração
Instituto Politécnico do Porto
São Mamede de Infesta, Portugal

José Fernando Oliveira
CEGI—INESC TEC; Department of Industrial Engineering and Management
Faculty of Engineering
University of Porto
Porto, Portugal

Carla Soares Geraldes
CeDRI-IPB; Department of Industrial Management
School of Technology and Management
Polytechnic Institute of Bragança
Bragança, Portugal

Samuel Moniz
Department of Mechanics
University of Coimbra
Coimbra, Portugal

Alberto Adrego Pinto
LIAAD—INESC TEC; Department of Mathematics
Faculty of Sciences
University of Porto
Porto, Portugal

ISSN 2194-1009 ISSN 2194-1017 (electronic)
Springer Proceedings in Mathematics & Statistics
ISBN 978-3-031-20787-7 ISBN 978-3-031-20788-4 (eBook)
https://doi.org/10.1007/978-3-031-20788-4

Mathematics Subject Classification: 90Bxx, 65XX, 91Axx, 49-XX, 90XX

This Springer imprint is published by the registered company Springer Nature Switzerland AG
The registered company address is: Gewerbestrasse 11, 6330 Cham, Switzerland

Preface

The XXI Congress of APDIO, the Portuguese Operational Research Society, took place on 7 and 8 November 2021, with more than 120 registered participants and around 80 papers submitted. This congress took place at a very particular historical moment. After the first waves of the COVID-19 pandemic, which had immobilized humanity, cancelling all gatherings, society was equipped with the first effective weapons against the pandemic, in particular vaccines, and the world was beginning to reopen. Science, in all its facets, had been massively called upon for this on a scale that humanity had never before experienced. It was in this context of hope that the APDIO congress took place, also celebrating the significant contribution of this discipline in fighting the pandemic. Thus, "Analytics for a Better World" was chosen as the congress theme.

The plenary talks were fully aligned with the congress theme. Manuel Matos spoke about uncertainty in decision processes. We know that uncertainty is the rule rather than the exception, but it has a special impact when human lives are at stake, in humanitarian logistics applications, or in disaster management. And this was the theme of the talk by Sérgio Guedes Silva, who leads the supply chain management team of the World Food Program (WFP), and who spoke about "The power of analytics in a Humanitarian Context". An equally sensitive mission was the one taken on by Vice Admiral Gouveia e Melo, who led the task-force in Portugal that managed the vaccination process against COVID-19. For the benefit of Portugal, the success of his work was overwhelming. Therefore, we wanted to hear from the Vice-Admiral the main challenges faced, and the lessons learnt in this process. The pandemic had unexpected impacts on various sectors of the economy, but logistics is one of the activities that is still most affected by the disruption suffered. Supply chain management is Pedro Amorim's area of work, and it was from this application that he reflected on how we can leverage the practical relevance of our research.

The works selected for this book have also, in different ways, contributed to a better world.

As shown in "A Biased Random-Key Genetic Algorithm for the Home Care Routing and Scheduling Problem: Exploring the Algorithm's Configuration Process", by Aguiar et al, one approach to reducing the expenses of the health system is to shift some of the undifferentiated care provision to social systems. Such is the case of home care, provided by social organizations which support the elderly and convalescent patients, contributing to reducing the demand for hospital care. But, and not less relevant, home care allows for more personalized treatment of these patients.

Forest fires cause incalculable damage to fauna and flora and lead to the death of people and financial damage in general. To avoid wildfire catastrophes is fundamental to detect fire ignitions in the early stages, which can be achieved by monitoring ignitions through sensors. "An Integer Programming Approach for Sensor Location in a Forest Fire Monitoring System", by Azevedo et al, deals with the decision of where to locate such sensors to maximize the coverage provided by them, taking into account different types of sensors, fire hazards, and technological and budget constraints.

The healthcare structure and quality of life of the population can be impacted by the efficiency of global pharmaceutical supply chains, through the price and the availability of medication. A bi-objective capacity allocation model that aims to generate cost-efficient and fair solutions is presented in "Capacity Allocation Incorporating Market Equity Concerns: A Pharmaceutical Supply Chain Case Study" by Bessa et al, minimising unfairness with a metric that accounts for drug shortages and maximising the economic value generated. Results suggest that a significant amount of unfairness can be tackled with little impact on economic targets.

The paper "The Shortest Path in Signed Graphs", by Costa et al, looks at shortest path problems in a signed graph. The shortest path in a signed graph can be used to understand how successive relations, even if distant, aaffect the dynamics of the network. Initially introduced to represent feelings among people belonging to the same social group, signed graphs were later used to model other systems, such as biological networks, international relations networks, risk management networks, i.e, systems a polarized environment is present and there is the willingness to consider it explicitly.

Having efficient manufacturing processes requires accurate failure detection to reduce equipment downtime. "The Break Point: A Machine Learning Approach to Web Breaks in Paper Mills", by Dias et al, presents a machine learning approach for predicting web breaks in tissue paper machines. Web breaks prediction plays a key role in ensuring product quality and sustainable use of energy, water, and other resources.

A real case addressing the optimization of fire brigades' rescue time is presented by Lima et al. "A Resectorization of Fire Brigades in the North of Portugal". Through a practical application of a Non-dominated Sorting Genetic Algorithm (NSGA-II) that performs the distribution of fire brigades into geographic areas, the authors propose a solution method to minimise the rescue time response in the case of forest fires, assuming the geographic and population characteristics of the areas and the fire brigades' capacity.

Firms can determine optimal operating policies based on agility and flexibility principles. In the work of Magalhães et al. "A Holistic Framework for Increasing

Agility in a Production Process", the authors address the complexities of real-world manufacturing systems by proposing a generic framework for increasing the agility of the production processes. Existing interdependencies between portfolio management, product complexity, equipment efficiency, and production planning decision-making are mapped into a set of methods that can enable flexibility.

To achieve high production efficiency, adequate additive manufacturing scheduling methodologies are important. In "Nesting and Scheduling for Additive Manufacturing: An Approach Considering Order Due Dates" by Nascimento et al, a constraint programming formulation is presented to solve a problem of simultaneously nesting irregular-shaped parts and scheduling additive manufacturing machines, balancing the fulfilment of order due dates with the usage of the machines' capacity.

The work proposed by Öztürk et al., "Developing a System for Sectorization: An Overview" also tackles the sectorization challenge by proposing a decision support system capable of solving various sectorization problems. These include real-life decision-making situations, such as school or health districting, logistic planning, maintenance operations or transportation. Several solution methods can be used depending on the structure of the problem to be solved.

In transporting hazardous materials or in cash collection it is important to find K dissimilar paths, that can work as alternatives or backups to one another in case of a failure in the network, while also minimizing the total cost. In "New Models for Finding *K* Short and Dissimilar Paths" by Pascoal et al., this bi-objective problem is modelled with integer linear programming and solved with ϵ-constraint method.

Healthcare services have critical delivery time windows, and consequently require maximum route optimisation. Pereira et al. in the chapter "Time Windows Vehicle Routing Problem to On-Time Transportation of Biological Products on Healthcare Centres", analysed and determined a set of vehicle routes to perform on-time transportation of biological products from seven local healthcare centres to a central hospital. Using a Vehicle Routing Problem with Pickups and Time Windows (VRPPTW) model, it was possible to improve the healthcare units' solution without further investments or reallocation of available resources

An effective communication could impact the incidence of an infectious disease, leading to new habits of the population, and at the same time, to improve the awareness of health centers' staff. In the chapter "The Role of Communication on the Spread of Dengue: An Optimal Control Simulation", the authors through a compartmental model related to vector-borne dengue disease carried out simulations, using distinct levels of communication by the authorities, aiming to show that an efficient channel of communication could save money to the Health System and could considerably decrease the number of infected individuals.

Optimizing the blood supply chain network is of uttermost importance for the rational use of a scarce and valuable resource: blood products. In "Towards an Optimized and Socio-Economic Blood Supply Chain Network", by Torrado et al, the design of a blood supply chain network is considered to support blood supply and demand and the geographical distribution for donors/patients according to the location of the healthcare facilities. In the design process, not only are costs minimized, but social aspects are also taken into consideration.

Road transportation is still a critical sector in terms of carbon dioxide emissions. To address this challenge Vaz et al., "A DEA Approach to Evaluate the Performance of the Electric Mobility Deployment in European Countries", describe a DEA approach to evaluate the electric mobility of the European countries, aiming at improving practices toward better road transport sustainability. Results indicate that most countries have the potential to improve road transport, for instance, by following the best practices adopted by the Netherlands or Sweden.

The economic point of view of retailers and manufacturers is addressed in "The Art of the Deal: Machine Learning Based Trade Promotion Evaluation" by Viana & Oliveira. A decision support system was developed to aid in the promotional planning process of olive oil and vegetable oil, to evaluate trade promotions from the point of view of the manufacturer, ensuring manufacturer margins. The simulation is powered by multiple gradient boosting machine models that estimate sales from the limited and unpolished data available.

This is a very thorough and rich set of examples where Operational Research applies analytics tools for a better world, demonstrating the vitality of the Operational Research community in Portugal.

Bragança, Portugal — João Paulo Almeida
Bragança, Portugal — Carla Soares Geraldes
São Mamede de Infesta, Portugal — Isabel Cristina Lopes
Coimbra, Portugal — Samuel Moniz
Porto, Portugal — José Fernando Oliveira
Porto, Portugal — Alberto Adrego Pinto

Contents

A Biased Random-Key Genetic Algorithm for the Home Care Routing and Scheduling Problem: Exploring the Algorithm's Configuration Process

Ana Raquel Aguiar, Tânia Ramos, and Maria Isabel Gomes

Abstract One approach to reduce the expenses of the health system is to shift some of the undifferentiated care provision to social systems. Such is the case of home care, provided by social organizations which support the elderly and convalescent patients, contributing to reduce the demand for hospital care. The problem is usually modeled as a VRPTW, which is a NP-hard problem and thus very complex to solve. This work develops a biased random-key genetic algorithm to design single-day caregiver routes for home visits. A particular emphasis is placed on the algorithm's configuration process, not frequently explored in the literature and a new methodology is suggested, based on the concept of performance profiles typically used for solver performance analysis. The performance profiles of configurations demonstrate the robustness of the metaheuristic, also providing a means for visually comparing the performance of different configurations and supporting the selection of one.

Keywords Home social care routing and scheduling problem · Biased random-key genetic algorithm · Metaheuristic · Performance profiles

1 Introduction

Over the last decades the aging of the population and other socio–demographic changes placed an additional burden on health systems. One of the directions to

A. R. Aguiar (✉) · T. Ramos
CEGIST, Instituto Superior Técnico, Universidade de Lisboa, Av. Rovisco Pais, 1049-001 Lisboa, Portugal
e-mail: a.raquel.aguiar@tecnico.ulisboa.pt

T. Ramos
e-mail: tania.p.ramos@tecnico.ulisboa.pt

M. I. Gomes
CMA-FCT, Faculdade de Ciências e Tecnologia, Universidade Nova de Lisboa, Monte da Caparica, 2829-516 Caparica Almada, Portugal
e-mail: mirg@fct.unl.pt

J. P. Almeida et al. (eds.), *Operational Research*, Springer Proceedings in Mathematics & Statistics 411,
https://doi.org/10.1007/978-3-031-20788-4_1

solve this issue is to improve the articulation of the health and social systems [5]. In particular, the provision of home care which encompasses activities of the daily living such as medication assistance, bathing, dressing, diaper changing, delivery of meals, among other services. Caregivers travel to the patients' residencies and provide these services. A great deal of attention has been placed on home care problems since they have the potential to lower healthcare costs by reducing the number of hospital admissions and respective lengths of stay [14]. Additionally, favouring home care promotes aging-in-place, thus reducing the demand for long-term residential care.

The home care provider is usually a social organization, facing challenges to achieve economies of scale and, consequently, having weaker prospects of becoming financially sustainable [20]. Therefore, designing efficient operational plans is important to achieve their social goals. The present work is focused on the routing and scheduling of home care visits and is motivated by a real case-study. Currently, the organization in study performs its operational planning manually and support planning tools are welcomed, in special, those able to quickly present a plan, preferably in less than 15 min.

The problem of designing routes to provide care has similarities with the well-known home healthcare routing and scheduling problem (HHCRSP) for which Di Mascolo et al. [7] present a general description as follows. A set of patients, represented as a set of nodes, spread over a territory, requiring different durations of care demanding different qualifications and which must be provided at the patient's residency. The care is provided by caregivers, managed by an organization, and distinguishable by different skills and availabilities. The characteristics of a patient are its availability, modeled through a time-window, along with a visit duration modelling the patient demand for care. To each arc connecting two nodes (patients) a weight is associated, such as a travel time or cost. In most cases, caregivers start and finish their routes at the headquarters of the organization, travel between patients mostly by car but optionally with other means of transportation, such as by bus, bicycle or walking. The HHCRSP then consists on allocating visits to caregivers and deciding visit starting times, while respecting several constraints, so as to optimize a certain objective over a given planning horizon.

In this work we introduce a problem inspired by the real case-study of a home care provider that presents some features not yet tackled by the literature. Instead of considering caregivers as the routing object, it considers teams of either one or two caregivers (designated from this point forward as single or double teams, respectively). In turn, these teams serve patients whose care requires one or two caregivers (hereinafter, designated as semi-dependent or bedridden patients, respectively). The aim is to design more efficient routing plans by introducing flexibility into the solution. Currently the planning is done separately for single and double teams, as this problem decomposition decreases the complexity for manual design. However, since the skills of the caregiver are homogeneous the bedridden patients can equally be served by synchronizing two single teams. Synchronization is understood as two single teams joining to provide a service to a patient requiring two caregivers (like a bedridden patient).

To solve the problem we apply a Biased Random-Key Genetic Algorithm (BRKGA), a metaheuristic first introduced by Gonçalves and Resende [12]. We develop a new decoder which selects the required number of each type of team, depending on the number of caregivers and cars available, as well as considering the proportion between patient types. As mentioned above, most applications of the HHCRSP consider caregivers as the routing object whereas we consider teams of one or two indistinguishable caregivers. This allows exploring the scheme of teams used to fulfill the demand. The efficiency of the algorithm is tested on adapted instances present in Mankowska et al. [17]. Being a metaheuristic, several parameters need to be tuned. This process is included in the configuration process, and it is frequently not shared on the literature. A configuration performance profile, based on solver performance profiles [8], is suggested to characterize different configurations of metaheuristics in general. The performance profiles provide information on the contributions of different configurations to a particular behaviour, for example, how two parameters contribute to increase run-time within their typical ranges of values applied.

The remainder of the paper is structured as follows, Sect. 2 exposes relevant related literature, while Sect. 3 defines the problem. Section 4 introduces the metaheuristic approach and describes the configuration process. Tuning results, as well as results comparing the metaheuristic with the implementation of the mathematical programming model in a commercial solver are analysed and discussed in Sect. 5. Finally, some conclusions and future work compose Sect. 6.

2 Related Literature

The problem at hand falls within the home healthcare routing and scheduling problem context. Recent reviews, such as Fikar and Hirsch [10], Cissé et al. [4] and Di Mascolo et al. [7] present and discuss the most common objectives, constraints and features characterizing the problem.

The objectives are usually route-related. Since in home care the working time of nurses/caregivers is the more expensive resource, the objectives are associated to caregiver's traveling, working and waiting times, as well as overtime [10]. After operational cost, a second major optimization objective found in the literature is related to quality of service and well-being, including the consideration of patient's preferences regarding the assignment of a specific caregiver, time-window violations and continuity of care, when the organizational preference is to allocate the minimal number of caregivers to any patient [7].

The solution methodologies employed are mostly of an heuristic nature. For instance, in spite of applying an exact solution method, in Gomes and Ramos [11] the problem is first decomposed and solved independently for different patient types. Several metaheuristics have been applied to solve single-day HHCRSPs, both regarding single-solution methods and population-based methods [4].

An Adaptive Variable Neighborhood Search is introduced in Mankowska et al. [17], where the authors introduce a new representation of a solution through a matrix. The initial solution is obtained through a greedy constructive heuristic, scheduling patients based on priority and suitability of caregivers allocated. This first solution is improved using 8 different search neighborhoods. The metaheuristic is considered adaptive because the neighborhoods to search are ordered by their potential to find good solutions. The same metaheuristic is used in Liu et al. [16], highlighting the increased difficulty of designing neighbourhoods for vehicle routing problems with interconnected routes, a consequence of the synchronization feature of many HHCR-SPs. The results presented by the author for tuning the metaheuristic are three tables with little readability. Rest et al. [19] suggest three Tabu-Search inspired methodologies. The initial solution is generated with a greedy heuristic, where all visits are inserted into the solution, first based on preference, then non-rejection and skill-level requirements and finally all services are inserted based on objective function cost. The initial solution may be infeasible. The neighborhood is based on shifting services between routes and the differences between the three methodologies concern the size of the neighborhood searched.

Examples of population based methods include Akjiratikarl et al. [2], who implement a Particle-Swarm Optimization (PSO)-based algorithm, which combines local improvement techniques. The algorithm explores the solution space globally with the evolutionary search characteristic of the PSO but also exploits neighbouring solutions through swap and insertion local search procedures. Initial solutions are generated based on the earliest time priority. The authors compare different variants of the algorithm and select the best based on ANOVA results. An ant-colony optimization algorithm combined with a memetic algorithm is the solution method implemented in Decerle et al. [6]. The parameters are said to have been set empirically and the author is warned about how the application of the algorithm to other problems may influence its performance. Nevertheless, they compare different algorithms by establishing a limit on run-time and assessing the objective function. Du et al. [9] introduce a genetic algorithm and a hybrid genetic algorithm, for which they provide the values of the parameters used without further comments, with a population size of 500, probabilities of crossover and mutation of 0.5 and the termination criteria being 3,000 generations. Another hybrid genetic algorithm is proposed in Shi et al. [22], where the authors consider fuzzy demands. The efficiency of the algorithm is tested by considering different features of the problem, for example, deterministic *versus* stochastic demand.

The only work we could find applying the BRKGA to the HHCRSP was presented in Krummer et al. [15], considering time-windows, synchronization and different caregiver qualifications. As previously clarified, our work differs by solving the problem where caregivers are homogeneous and where the most relevant problems is how to allocate them to two different types of teams. The latter problem is more closely related to the social care context and has not been explored in the literature.

One of the common characteristics of the above mentioned works is that they do not detail the process used for configuring the applied the metaheuristics. An exception is Akjiratikarl et al. [2], who share that they use the Taguchi design of experiments and use it to set the balance between global exploration and local exploitation of the algorithm. As far as the authors know, performance profiles have not been used to analyse metaheuristic configurations. These profiles were introduce to compare solvers on a set of problems with different sizes [8]. We propose to look at the configurations as though they were solvers being compared with each other.

3 Problem Description: A Home Care Routing and Scheduling Problem

A set of patients requires one daily task, $i \in N_C$, associated to different durations of care, W_i, which must be provided at the patient's residency. The patients are available at a pre-defined time-window, $[e_i, l_i]$, and the care they request may require the presence of one or two caregivers, R_i^C. Tasks requiring one caregiver are usually placed by more independent patients and are thus denominated semi-dependent, $i \in N_S$ whereas tasks requiring two caregivers are frequently placed by patients whose mobility is compromised and are, therefore, denominated bedridden, $i \in N_B$. A task of a bedridden patient requires services which are more physically demanding such as transferring patients from the bed to living room or giving them baths while lying on the bed. All tasks are classified as either semi-dependent or bedridden: $N_C = N_B \dot{\cup} N_S$.

The care is provided by a homogeneous set of A^C caregivers, with a maximum shift length, H. They must start and finish their routes at the day-care center. The caregivers travel in cars and the number of available cars, Q, usually lower than the number of caregivers. The service provider organization assigns the caregivers to teams, $k \in V$, of either one or two caregivers, denominated single, $k \in V_S$, and double, $k \in V_D$, teams respectively. Set $V = V_S \dot{\cup} V_D$, where $|V_S| = \min\{Q, A^C\}$ and $|V_D| = \min\{Q, \lfloor \frac{A^C}{2} \rfloor\}$. The number of teams on the solution must be less than or equal to the number of vehicles. Equally, the sum of the number of caregivers in each team (M_k) on the route plan must not exceed A^C.

Currently route design is performed independently for single and double teams, with the number of teams of each type fixed a priori. Single teams serve semi-dependent patient while bedridden patients are served by double teams, exclusively. We aim to introduce flexibility into route design by allowing bedridden patients to be served by two synchronized single teams and by allowing double teams to serve semi-dependent patients. Synchronization reduces the need for double teams. At the limit, all bedridden services could be performed by synchronized single teams. However this is frequently impossible due to the constraint imposed by the number of vehicles limit. Additionally, shifting some of the semi-dependent demand to a double teams may allow a reduction in the number of caregivers required to answer the same demand for care.

The problem then consists on deciding how to allocate the tasks to teams, how many teams of each type to choose, when to use synchronization and the sequence by which visits should be performed. The selection of teams of each type is influenced by factors such as the proportion of demand for each service type (semi-dependent vs. bedridden) or the geographical locations of services. The objective function can be mathematically formulated as given by Eq. 1. Parameter T_{ij} stands for the traveling time between two locations and decision variable x_{ijk} is equal to 1 if team k travels from node i to node j, and 0 otherwise. The problem is mathematically formulated in Aguiar et al. [1].

$$\min_{x} \sum_{k \in V} \sum_{i,j \in N} (T_{ij} + W_i) M_k x_{ijk} \tag{1}$$

Notice that if a semi-dependent service is allocated to a double team the objective function will be penalized as the service duration is multiplied by the number of caregivers providing the task.

4 Methodology

We propose a BRKGA to solve this variant of the HHCRSP problem. The general framework of the algorithm is introduced in Sect. 4.1, followed by the description of the steps in the decoder (Sect. 4.2) and finally Sect. 4.3 highlights relevant aspects of the methodology configuration, namely the control flow and tuning process.

4.1 The Biased Random-Key Genetic Algorithm

Biased Random-Key Genetic Algorithms are frameworks to build heuristics to solve combinatorial optimization problems, which can be applied to a wide range of problems. Being an evolutionary algorithm, in each generation a population evolves by producing offspring and by introducing mutants. Each individual in a population represents a solution. In other words, for each individual there is a corresponding chromosome, composed of a vector of genes, encoding the solution. The genes take on values, called alleles, which vary between 0 and 1, the so-called random keys. This encoding of solutions was introduced by Bean [3] with the Random-Key Genetic Algorithms (RKGA). A decoder maps the vector of random keys to a solution of the optimization problem and computes its fitness [13].

Algorithm 1 : The framework of the BRKGA

Require: p, p_e, p_m, ρ_e
1: Generate initial population with p vectors of random keys
2: **while** Stopping criteria not met **do**
3:     **Decode** each solution in the population
4:     Sort solutions by fitness
5:     Classify solutions as Elite or Non-Elite
6:     Copy Elite solutions for next generation
7:     Generate mutants for next generation
8:     Generate offspring and add it to the next generation
9: **end while**
10: **return** Decoded Best Solution

An overview of the BRKGA framework is presented in Algorithm 1. The method starts by generating an initial population of p individuals, with each allele determined randomly. Then, the fitness for each individual is calculated using the decoder and the population is partitioned into the Elite set, composed of the p_e individuals with the best fitness ($p_e \leq \frac{p}{2}$) and the other $p - p_e$ individuals make up the Non-Elite set. The Elite solutions are copied to the next generation followed by the addition of the p_m mutant solutions (vectors with all the alleles randomly generated). The copying of the Elite individuals implements the concept of the survival of the fittest. The remainder $p - p_e - p_m$ individuals of the population are offspring resulting from parametrized uniform crossover (introduced by Spears & De Jong [24]). Each offspring inherits the allele from the Elite parent with probability ρ_e. This is the point where BRKGA differs from the RKGA. While in RKGA the parents are chosen randomly among all individuals in a population, in BRKGA one parent is selected from the Elite individuals and the other from the Non-Elite individuals. The population evolves until meeting a stopping criteria. One very interesting aspect of this metaheuristic is the separation between the evolutionary part (which is problem independent) and the fitness assessment (which is problem dependent)—the decoder [12]. Depending on the decoder, the methodology may even be considered a hybrid metaheuristic [18].

4.2 *The Decoder*

A decoder is usually a deterministic heuristic that translates the Random Key vector, i.e. the chromosome, into a solution as fast as possible. Therefore, decoders are usually constructive heuristics [18], as the one proposed in this work. A chromosome is a vector of n random keys in which each position is associated to a task. The decoder then takes each task (i.e. each gene) and assigns it to the route(s) with the minimal insertion cost, while complying with problem constraints. Our decoder comprises essentially two algorithms, a main one which receives a task, assesses its type and decides where to insert it by calling the second algorithm, the insertion cost calculation.

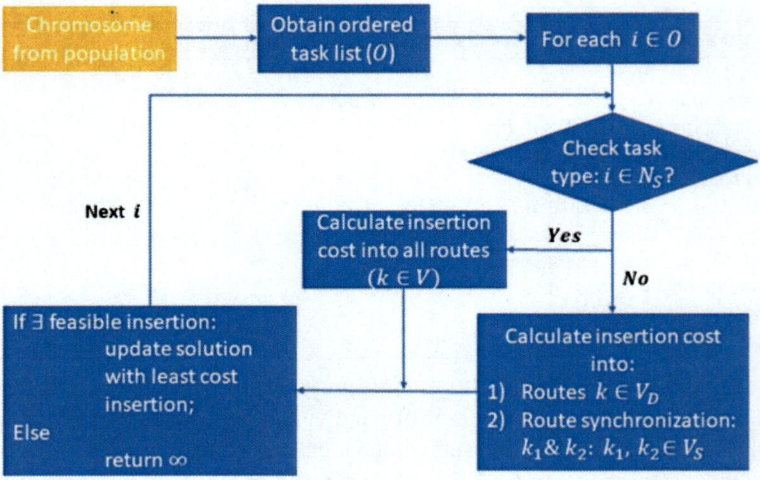

Fig. 1 Decoder scheme

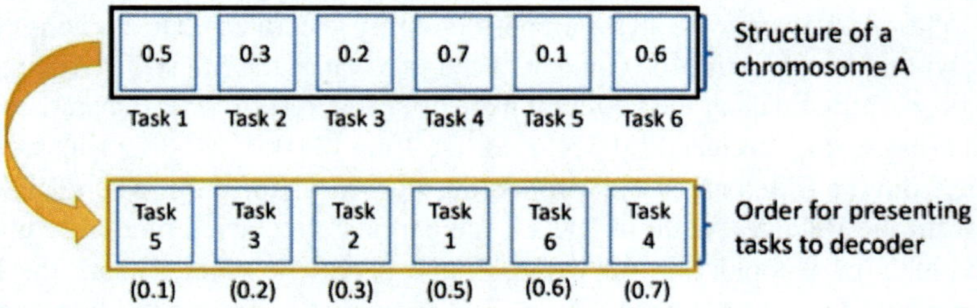

Fig. 2 Chromosome representation

Figure 1 presents a scheme of the main steps of the decoder. The decoder receives the chromosome which is made of gene-allele pairs and sorts by ascending order the tasks (genes) in accordance with the values of the Random Keys (alleles). For instance, Fig. 2 shows a chromosome for a problem with 6 tasks, i.e., the chromosome is a vector of alleles, say $<0.5, 0.3, 0.2, 0.7, 0.1, 0.6>$, where each position corresponds to a task, say < task1, task2, …, task6>. The decoder firstly reorders it according to the value of the alleles. The ordered vector will then be $<0.1, 0.2, 0.3, 0.5, 0.6, 0.7>$ that translate into $O = <$ task5, task3, task2, task1, task6, task4 $>$. Each task in O must then be inserted into the solution. For semi-dependent services ($i \in N_S$) the insertion cost into each route is calculated and, if it exists, the least cost insertion is chosen and the solution updated. In turn, for bedridden services ($i \in N_B$), the insertion tests comprise only double routes and a pair of single routes, implying synchronization. When the task cannot be inserted into the solution, the solution is infeasible and the decoder returns ∞ as its fitness value. The algorithm for insertion cost into a route also verifies many of the problem constraints.

4.3 Configuration Methodology

The general components in evolutionary algorithms are the mutation, crossover and selection operators. To some extent, these components can exist independently from the rest of the algorithm and are called by it in sequence. Different implementations may call the operators in different orders (e.g. crossover may be done before mutation or the other way around). Each component may also be characterized by a set of parameters, whose values are set by the algorithm designer. The configuration of a metaheuristic is defined as the specific control flow and the specific set of parameter value it uses [21].

The configuration of an algorithm may have a great influence on its performance, both in terms of solution quality and of search efficiency. However, there is no optimal configuration that solves all problems and all their instances [23] and therefore the configuration process is often a difficult and time-consuming phase. After making the solution representation choices, the control flow is set through trial-and-error and the values for the parameters are obtained empirically. Furthermore, the control flow and parameter values influence each other, which renders the configuration process even more difficult. Regarding control flow, the original BRKGA randomly generates the initial population (Elite and non-elite). In the current work, a different approach is followed to create the initial Elite population. This additional step is needed since, for larger instances, the metaheuristic does not easily find feasible solutions when tasks are ordered from a randomly generated chromosome. An initial solution is created attributing equally spaced values between 0 and 1 to the tasks, based on earliest times of tasks visiting time window (e_i). For instance, suppose a problem with 6 tasks where the time window earliest time, in an ascending order, renders the sequence: task 3, task 4, task 2, task 5, task 1, task 6. The corresponding chromosome is the vector $< 0.8, 0.4, 0, 0.2, 0.6, 1 >$. The initial elite population will then be formed by replicas of this vector.

Parameter tuning was conducted in two phases. The first phase was dedicated to parameters related with the composition of the population, namely, the values of a, the factor that determines the size of the chromosome population as a function of the number of tasks ($a * n$), p_e, the percentage of elite individuals in the population and p_m, the percentage of mutant individuals in the population. In order to do so, a full-factorial design of experiment was followed. Since the BRKGA metaheuristic is stochastic, each combination was run for different seeds to assess the variability of results. After choosing one combination, the remaining parameters, ρ_e, the probability of inheriting an allele from the elite parent and g, the number of iterations, were tuned (the second phase).

The evaluation of different values for the parameters is done through performance profiles, similar to those proposed by Dolan and Moré [8]. They present the advantage of allowing the comparison of problems with different sizes. After deciding which are the relevant measures (in our case the objective function and the run-time) the benchmark results are obtained by running the metaheuristic with the configuration c, from the n_C configurations established in the full-factorial design, for a set of n_Π

problems of different sizes. For each problem and configuration, results are obtained for n_S seeds.

The analysis of results is performed on ratios. Let $t_{\pi,c,s}$ denote the measured performance of configuration $c \in C$, on problem $\pi \in \Pi$, with seed $s \in S$, for a specific metric and, assuming that $t_{\pi,c,s} > 0$ and that lower values of $t_{\pi,c,s}$ indicate a better performance, the performance ratios are then given by Eq. (2),

$$r_{\pi,c} = \frac{\bar{t}_{\pi,c}}{min\{t_{\pi,c,s} : c \in C, s \in S\}} \tag{2}$$

where

$$\bar{t}_{n,c} = \frac{1}{n_S} \sum_{s} t_{\pi,c,s} \; and \; n_S = |S|.$$

The expression $min\{t_{\pi,c,s} : c \in C, s \in S\}$ represents the best known objective (BKO) for problem π. Performance profile of a configuration $c \in C$ (combination of parameter values) is the empirical cumulative distribution function for the ratios $r_{\pi,c}$. It is defined by Eq. (3),

$$\delta_c(\tau) = \frac{1}{n_\Pi} |\{\pi \in \Pi : r_{\pi,c} \leq \tau\}| \tag{3}$$

with $\delta_c(\tau)$ being the probability that configuration c is within a factor τ of the best performance over all configuration of all problems $\pi \in \Pi$. As such, each result for a problem is only compared with results for that problem when determining $r_{\pi,c}$.

5 Results and Remarks

The metaheuristic was implemented using the programming language python, version 3.7.4, and run on a computer with an Intel(R) Core(TM) i5-8500 CPU @ 3.00 GHz 3.00 GHz processor and 8 GB of RAM. The instances used for the benchmark tests were adapted from Mankowska et al. [17]. In these instances, shared visits include both synchronizations and services with a given precedence relation, i.e., some tasks can only be performed if another task has been performed previously. In the current work, it is assumed that all shared visits need to be serviced by two caregivers (either on a double team or by the synchronization of two single teams). Also, the traveling time between nodes is reduced to a third, representing more closely the conditions of the partner organization. From the set of problems we use the 10 available instances of sizes 10, 25 and 50, adding up to 30 different instances (or problems) of 3 different sizes. The problem size indicates the number of tasks in that instance. After completing the configuration process, the instances are also solved with the chosen configuration in Sect. 5.3, for performance comparison with the exact method.

Table 1 Parameters of BRKGA

Identifier	Name	Range
p	Population Size	$a * n$
a	Factor a	$\{2, 5, 10\}$
p_e	Elite Population Size	$\in]0.1p, 0.3p]$
p_m	Mutant Population Size	$\in]0.1p, 0.3p]$
ρ_e	Probability of inheriting allele from fittest parent	$\in]0.5, 0.8]$
g	No. Generations	100–1000

The configuration performance profiles can be used both for control flow and tuning configuration alternatives, but in this work the methodology is tested for the tuning phase. The first phase of tuning concerned the parameters associated with the population, precisely p, or factor a, to be multiplied by n, p_e and p_m. The range of values tested for the parameters are made available on Table 1. For the first stage of tuning, the tuning of population-related parameters, the 3 parameters, a, p_e and p_m, assume 3 different values, leading to 27 different combinations tested, henceforth denominated configurations. In turn, each configuration is run for 5 different seeds. Parameter p is established by multiplying factor a by n, the number of tasks in each instance. Therefore, each configuration in the first stage is identified as $(a\_p_e\_p_m)$. The values for p_e and p_m are codified as $10\% = 1$, $20\% = 2$ and $30\% = 3$. In the first stage the values for ρ_e and g are fixed to 0.7 and 500, respectively. Although not shown, the choice of 500 generations was further validated concerning algorithm convergence. Then, in the second phase, parameters ρ_e and g are tuned, where each configuration is represented as (ρ_e_g). The two measures chosen to assess the configurations are the objective function value, assessing the quality of the solution and the run-time, assessing the efficiency of the metaheuristic.

5.1 *First Phase*

Figure 3 shows the performance profiles for the Objective Function (OF) test runs. Each line represents a different combination of values for the parameters. The three colors indicate the value for parameter a in each of the configurations (a different configuration). The lines represent $\delta_c(\tau)$ for the performance measure OF. One of the most relevant information to retrieve from the performance profiles is the point at $\tau = 1$, as it represents the probability of that configuration having the best performance (several configurations can have the same performance). In Fig. 3 the configuration guaranteeing the highest percentage of problems solved attaining the BKO has $a = 10$ and the guaranteed percentage is around 54%. Looking at Fig. 4, that configuration is identified as (10_3_3). Another relevant conclusion to retrieve

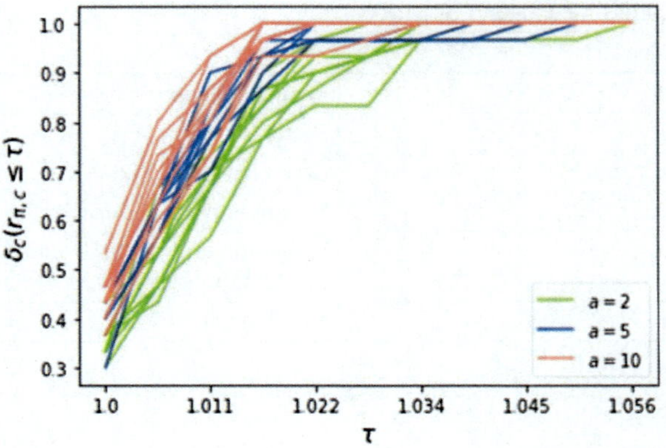

Fig. 3 Performance profile for OF benchmark tests ($a = 2, 5, 10$)

from Fig. 3 is the value $max(\tau) = 1.056$, indicating that for all tested configurations, the variation between the BKO for a problem and the OF of a solution obtained with any configuration is 5.6%. The previous result shows that the BRKGA is a robust method, as robust metaheuristics will have smaller variations of the OF associated to small changes to their parameters.

Overall, Fig. 3 shows clearly that $a = 10$ presents the best performance and $a = 2$ presents the worst performance. Although this is a common behavior in metaheuristcs, this figure also shows that for some configurations, $a = 5$ presents competitive results.

Setting the parameters p_e and p_m is less straightforward. There is no configuration that works well for all problems, as displayed in Fig. 4. However, configuration ($p_e = 3\_\ p_m = 3$) works well for a equal to 5 and 10. In both cases, the solutions for all problems are within 2% of the BKO for those instances (yellow line in Fig. 4). Moreover, the configuration (a_3_3), the yellowish line, dominates almost always the other remaining parameter configurations. (a_3_3) dominates in at least some parts of the respective charts, for example, in all charts the yellow line dominates for ($\tau = 1$). In short, one can say that this configuration is the most likely to attain the BKO among all configurations studied.

Actually, configuration (a_3_3) dominates all others in chart $a = 10$ and, in chart $a = 5$, (5_3_3) is only partially dominated by (5_1_3), (5_2_3) and (5_2_2). Configuration (5_1_3) has one of the greatest differences of performances for τ around 1% but is a much less robust configuration than any of the other three configurations. Also, configuration (5_2_3) always dominates or equals the performance of (5_2_2). The performance profiles of configurations (5_2_3) and (5_3_3) are very similar but placing the focus on (10_2_3) and (10_3_3), paints a different picture. Configuration (10_3_3) dominates all other profiles and has more than 50% chance of attaining BKO. This characterization of results for the OF reinforced by the results for the run-time measure, presented bellow, lead to the selection of configuration (5_3_3).

Concerning run-time (RT), Fig. 5 shows that there is one configuration with $a = 2$ that has a probability of having the best performance in run-time in more than 80%

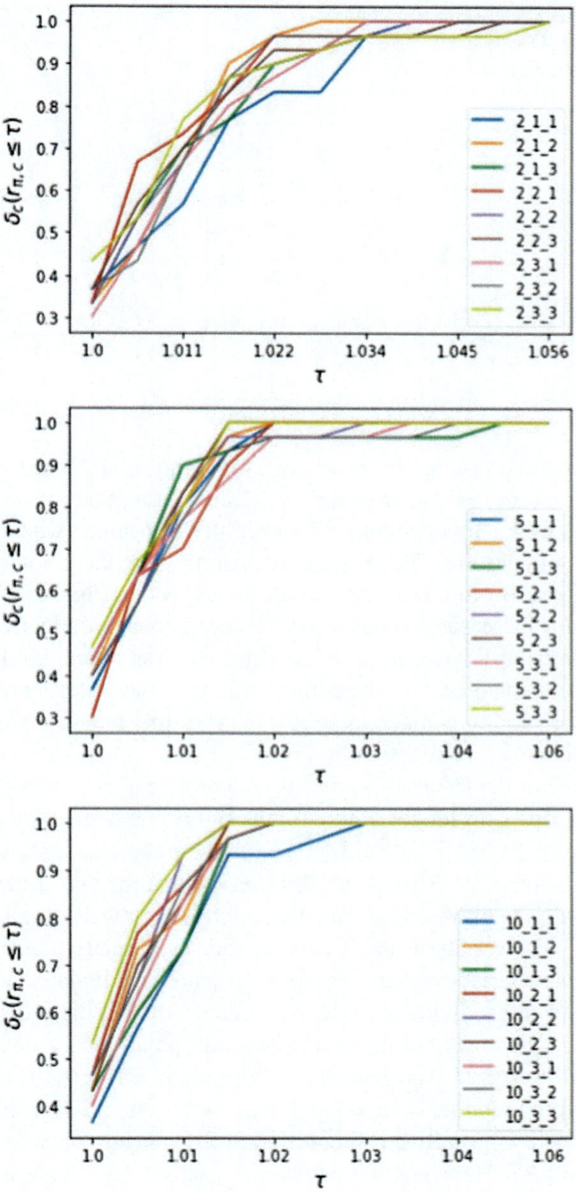

Fig. 4 Performance profiles for OF, aggregated by a, from top to bottom $a = 2$, $a = 5$ and $a = 10$. Legends state, from left to right, parameters $a = 2, 5, 10$, p_e, where $1 = 10\%$, $2 = 20\%$, $3 = 30\%$, and the same encoding for p_m

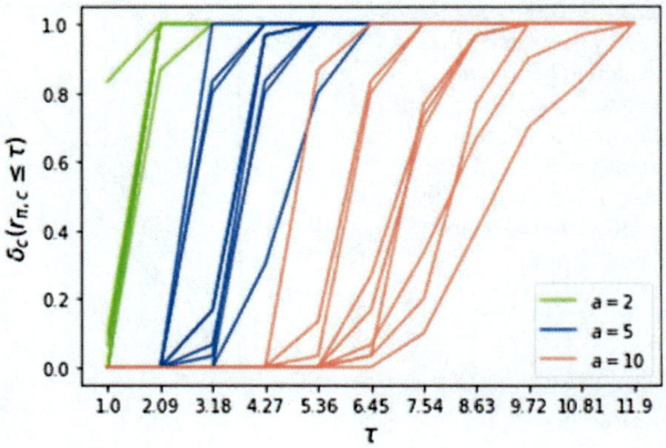

Fig. 5 Performance profile for RT benchmark tests ($a = 2, 5, 10$)

of the problems. For the lines regarding $a = 10$, the profile leaves the horizontal line $\delta_c(\tau) = 0$ at around $\tau = 4.3$, indicating that the best configuration with $a = 10$ takes 4.3 times more time to run the algorithm (when compared to the best tested configuration). The main conclusion is that the greater the value of a the longer it takes to run a test, as it would be expected. Figure 5 also depicts that as the value of a increases so does the difference between the performance of configurations in terms of RT within the same value of a (as expected). Remember that the values of τ compare each of the configurations to the fastest configuration. Also the range of the τ factor is much larger for the run-time measurement ($max(\tau) = 11.9$) than for the objective function ($max(\tau) = 1.056$).

For the values of p_e and p_m the reverse happens (Fig. 6): the smaller the values of p_e or p_m the longer it takes for the algorithm to run. Figure 6 shows that configurations (*a*_3_3), (*a*_3_2) and (*a*_2_3) preform the best, among the three values tested for parameter a. Also, it can be observed that the performance patterns are very similar with the previous combinations performing well for all values of a (Fig. 6).

The variations in RT performance are directly related to interactions between the characteristics of the population, namely its dimension and the numbers of mutants and elite individuals, with the decoder, which is the heaviest procedure in terms of RT. The greater the value of a the larger the population and the more times the decoder has to be applied. The greater the Elite population (p_e), the fewer the decoder is called, as more individuals are copied to the next generation with their fitness already known. There is usually a difficult trade-off between guaranteeing the quality of solutions and RT. However, the performance profiles of configuration (5_3_3) attest a most agreeable compromise.

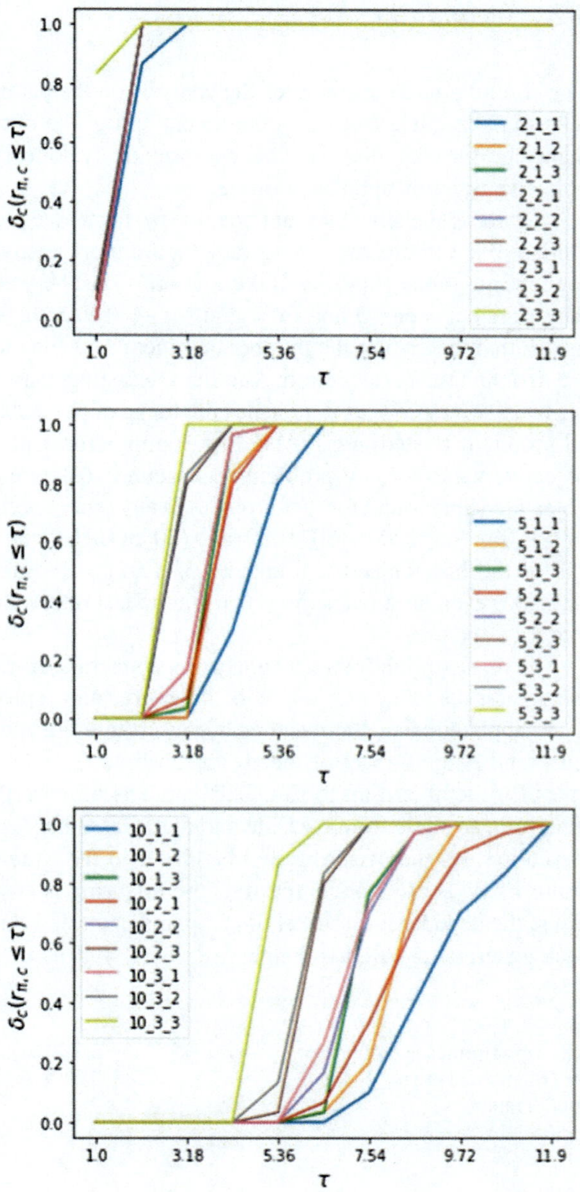

Fig. 6 Performance profiles for RT, aggregated by a, top to bottom, $a = 2$, $a = 5$ and $a = 10$. Legends state, from left to right, parameters $a = 2, 5, 10$, p_e, where $1 = 10\%$, $2 = 20\%$, $3 = 30\%$, and the same encoding for p_m

5.2 Second Phase

The second phase concentrates the analysis on the parameters ρ_e, the probability of inheriting an allele from the elite parent and g, the number of generations. Performance profiles, similarly to what was previously done, report on the measures of the objective function and the run-time.

Regarding the objective function, the performance profiles are displayed in Fig. 7. Unlike in the first phase, some patterns are more easily spotted. Looking at $\tau = 1$, for configurations with $g = 100$ the profiles start between 30 and 40%, for $g = 500$ they start between 40 and 50% and for $g = 1000$ between around 56–78%. This means that the greater the number of generations the more likely the configurations are to obtain the best solution. Another interesting analysis is to assess the value of τ for which $\delta_c(\tau) = 1$, as it indicates the range of percentage differences within which all problems (tested) are solved. For example, for configuration (50_100) the worst objective value for any problem was around 10% worst ($\tau = 1.096$) than the best objective value found for that problem in any other configuration. For configurations (80_1000), (65_1000), (50_1000) and (80_500) the worst objective value was around 2% of the best objective found, as $\delta_c(1.02) = 1$. Interestingly, the configuration (80_500) even dominates (50_1000) in the last interval before hitting the horizontal line $\delta_c(\tau) = 1$.

The most straightforward conclusion concerns run-time, see Fig. 8. As expected, the greater the value of g the worst the performance profile, as greater values imply more application of the decoder because there are more generations to assess. A sub-trend is the impact of the ρ_e on run-time. The smaller the ρ_e the least time it takes for the algorithm to run. This happens because the higher the probability of inheriting an allele from the Elite parent the less likely is the decoder to stop building a solution. As the offspring will be closer to the Elite parent that, as stated in the control flow subsection, is feasible, the offspring is also more likely to be feasible. Thus, the decoder is less likely to stop and assessing the fitness of all individuals in each generations will take longer.

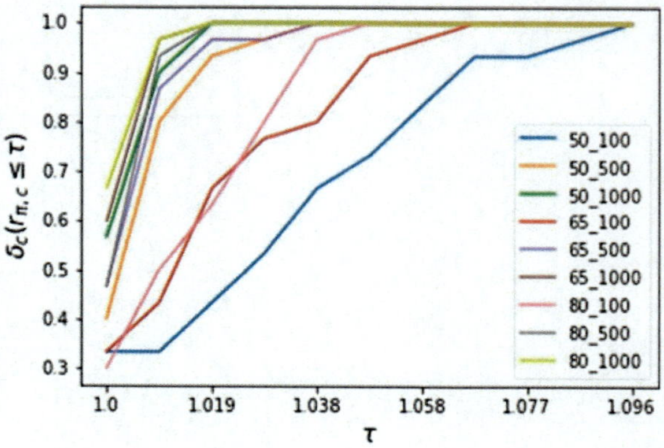

Fig. 7 Performance profile for OF measure of (ρ_e_g) configurations

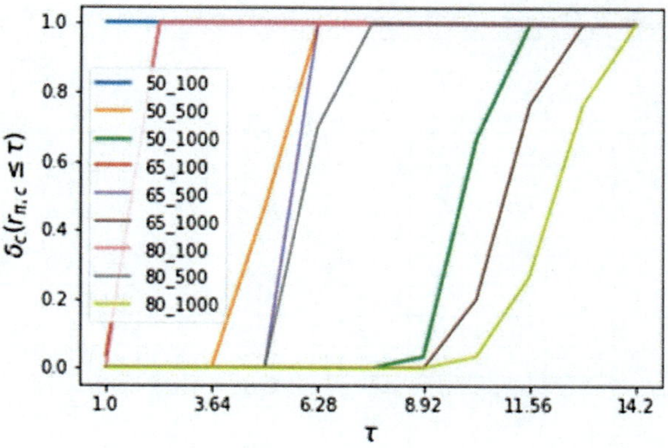

Fig. 8 Performance profile for RT measure of (ρ_e_g) configurations

The performance profiles provide relevant insights concerning the relative performance of different configurations of the BRKGA. They do so in a visual and rather intuitive manner. The chosen values of parameters (ρ_e, g) were (80_500) even though the maximum absolute run-time for configurations with $g = 1000$ was around 10 min, less than what would be the maximum accepted run-time. Since the performance profile for (80_500) regarding the objective function is just slightly worst than those with $g = 1000$, but considerably faster, the former configuration was the selected to compare the BRKGA with the exact method, present in the following subsection.

We highlight, however, that in the first phase the variation of the objective function was at most 5.6%, whereas for the second phase, the objective function is at most 9.6%. This is likely due to the fact that 100 generations is too tight for the algorithm to converge for the configuration (5_3_3).

5.3 *Exact Method Versus BRKGA*

This section compares the final configuration of the BRKGA ($a = 5$, $p_e = 3$, $p_m = 3$, $\rho_e = 80$, $g = 500$) with the implementation of the model of the HHCRSP described in the problem description section. The model was implemented using the commercially available software GAMS Studio 34.2 and solved with CPLEX version 20.1 on a workstation with a Intel(R) Core(TM) i9-10850K CPU @ 3.60 GHz 3.60 GHz processor with 128 GB of RAM. The analysis presented concerns the solutions obtained for all ten instances of sizes 10, 25, 50.

Table 2 shows the solutions obtained using the exact solution method and the BRKGA, measured by the objective function values (OF, in minutes) and the performance in run-time (RT, in seconds). As the BRKGA has a stochastic nature, it was run for 5 different seeds and the maximum and minimum values for each performance measure are also displayed in the table. Nevertheless, performance measures are compared using the average value obtained for the five tests performed with the

Table 2 Comparison of performances of the BRKGA and the exact method

		BRKGA						MIP			Comparison		
		Objective function			Run-time			OF	RT	gap	ΔOF(%)		ΔRT_{avg}(%)
Size	Id	max	min	avg	max	min	avg				avg	min	
10	1	412.5	403.2	406.7	2.6	2.3	2.4	365.5	0.8	0	−11.3	−10.3	−199.8
	2	348.0	348.0	348.0	2.4	2.3	2.4	309.0	0.4	0	−12.6	−12.6	−490.5
	3	505.9	498.0	502.7	2.4	2.4	2.4	404.2	4.2	0	−24.4	−23.2	42.4
	4	393.3	392.4	392.8	3.2	3.0	3.1	363.2	12.7	0	−8.1	−8.0	75.6
	5	321.1	321.1	321.1	3.1	2.9	3.0	279.4	2.4	0	−14.9	−14.9	−25.0
	6	415.7	412.2	414.3	3.2	3.1	3.2	381.8	0.2	0	−8.5	−8.0	−1491.0
	7	368.6	364.0	364.9	3.2	3.0	3.1	350.6	1.3	0	−4.1	−3.8	−138.2
	8	391.3	391.3	391.3	2.5	2.3	2.4	334.3	1	0	−17.1	−17.1	−137.6
	9	446.1	446.1	446.1	3.0	2.3	2.6	398.7	1.3	0	−11.9	−11.9	−97.2
	10	390.0	390.0	390.0	3.4	3.3	3.3	368.4	0.6	0	−5.9	−5.9	−457.0
Average		399.2	396.6	397.8	2.9	2.7	2.8	355.5	2.5	0.0	−11.9	−11.6	−291.8
25	1	810.8	798.3	803.3	22.9	20.4	21.3	742.7	13094	0	−8.2	−7.5	99.8
	2	719.5	696.9	705.7	28.3	23.3	24.5	651.2	11301	5.5	−8.4	−7.0	99.8
	3	827.8	823.4	824.5	20.7	19.0	19.9	766.1	918	0	−7.6	−7.5	97.8
	4	849.2	839.3	843.8	26.5	25.5	26.1	825.2	21346	0	−2.3	−1.7	99.9
	5	650.4	614.4	633.6	30.3	28.8	29.5	604.9	7546	4.6	−4.7	-1.6	99.6
	6	948.0	919.9	930.6	27.4	20.6	25.3	888.7	20887	4.9	−4.7	−3.5	99.9
	7	705.7	662.6	686.7	20.7	18.6	19.1	620.1	2407	0	−10.7	−6.9	99.2
	8	753.3	742.9	746.4	24.1	19.0	20.3	688.6	20516	0	−8.4	−7.9	99.9
	9	1015.4	915.3	940.4	22.8	19.2	20.8	852.9	48415	5.0	−10.3	−7.3	100.0
	10	854.9	820.7	834.7	26.1	21.3	22.7	798.1	41292	5.1	−4.6	−2.8	99.9
Average		813.5	783.4	795.0	25.0	21.6	22.9	743.9	18772.0	2.5	−7.0	−5.4	99.6
50	1	1402.9	1380.0	1390.3	192.2	180.8	187.4	–	–	–	–	–	–
	2	1214.5	1156.5	1180.8	187.6	153.7	174.5	–	–	–	–	–	–
	3	1480.0	1468.0	1475.8	185.8	169.0	176.8	–	–	–	–	–	–
	4	1554.9	1524.1	1545.7	201.3	170.0	182.6	–	–	–	–	–	–
	5	1208.2	1170.5	1184.4	204.7	195.1	201.2	–	–	–	–	–	–
	6	1692.8	1675.3	1684.5	186.9	169.3	178.5	–	–	–	–	–	–
	7	1283.9	1270.4	1278.4	198.7	186.2	192.8	–	–	–	–	–	–
	8	1313.1	1286.8	1299.2	236.5	206.0	220.3	–	–	–	–	–	–
	9	1643.3	1615.6	1625.7	157.2	151.7	153.9	–	–	–	–	–	–
	10	1430.9	1418.8	1425.5	206.5	201.7	203.7	–	–	–	–	–	–
Average		1422.5	1396.6	1409.0	195.7	178.3	187.2	–	–	–	–	–	–

BRKGA for each instance. For the size-10 instances, the objective function for the BRKGA is on average 11.9% worse than the exact solution method. Even though the run-time is considerably worse, most of the times, it is within the acceptable time for a company to obtain a solution. In turn, for the size-25 instances, the BRKGA objective function is about 7% worse, on average, than that of the exact solution method. However, for this instance size the run-time performances are much different. The MIP model would sometimes take a long time to even obtain a solution (even longer to prove optimally), running out of memory for some instances on a computer with more computing power than the one mentioned above. On average the run time was around 19 000 s. The BRKGA is able to obtain solutions in 30 s or less, allowing the algorithm to be run several times and for the best solution to be presented. If we were to consider this methodology (obtaining results for 5 different seeds and selecting the best as output), the BRKGA would perform around 5.4% worse on average, taking at most 150 s (30 s × 5 seeds). As expected, as the size of the instances increases, so does the relative performance of the BRKGA. The variation of the objective function between the two methods decreases from 11.9% to 7%. It is expected that this trend would continue. Nevertheless, for future work, local search should be explored as a means for improving the performance of the metaheuristic algorithm and closing the gap with the exact method. For instances of size 50 the exact method was allowed to run for 2 h but no solution was found. Therefore, even if the solutions of the BRKGA have a considerable gap, the algorithm is able to produce a feasible solution.

6 Conclusion and Future Work

In this work we applied a BRKGA to solve the home care routing and scheduling problem, where teams are the routing object. These teams are formed by either one or two caregivers and serve patients requiring one or two caregivers. The algorithm decides on how many teams of each type to use, as well as if patients requiring two caregivers should be served by one team of two caregivers or two (synchronized) teams of one caregiver.

The tuning of the metaheuristic was done using configuration performance profiles, which presents as main advantages the exploration of the performance of different configurations over problems of different sizes and to understand visually how the configurations compare to one another. Therefore, this approach is particularly relevant when evaluating how general is the method and in communicating this information visually, avoiding large tables whose analysis is usually challenging. For the range of tested values, factor a has the greatest impact on the algorithm performance in the first phase while the number of generations g had the greatest impact on the second phase. Factor a determines the number of individuals in a population, and the larger the number of individuals in the population, the longer the algorithm takes to run and the better is the solution. A similar performance behavior is verified for g. The configuration performance profiles also allowed a clear observation of a sub-trend regarding the probability of inheriting an allele from the fittest parent ρ_e. The higher the value of ρ_e the longer the algorithm takes to run and the better the solutions yielded. The BRKGA implemented produces solutions for problems of sizes 10, 25 and 50 in acceptable time, unlike the exact method, which could only yield solutions for problems up to size-25 taking on average around 19k seconds, not guaranteeing optimally for all solutions and some tests run out of memory. The relative performance of the BRKGA improves as the size of the instance increases.

Regarding the future work, we point directions concerning both the problem, the application of the metaheuristic and the usage of performance profiles. The problem could be further explored through the study of different objective functions, and multi-objective analysis by including workload balance objectives. Also, instead of adapting previously existing instances, tailored ones should be created as well as methodologies for solving a multi-period problem. Finally, uncertainty on service duration should be accounted for, as it is present in the context of the problem. Highlighting directions for enhancing the metaheuristic, we suggest the exploration of different procedures for initializing the population, which could impact the convergence of the metaheuristic. Finally, the methodology presented could benefit from the hybridization with a local search technique to check for improvements in the vicinity of each solution in the final population. Concerning the future usage of performance profiles, instead of considering separate sets of parameters in two stages, all the full-factorial experiment should include all parameters. Considering all parameters in the would allow for conduction of a more robust analysis.

Acknowledgements This work is funded by national funds through the FCT—Fundação para a Ciência e a Tecnologia, I.P., under the scope of projects UIDB/00297/2020 (Center for Mathematics and Applications), and UIDB/00097/2020 (CEGIST), along with the PhD grant SFRH/BD/148773/2019.

References

1. Aguiar, A., Ramos, T., Gomes, M.: The home care routing and scheduling problem with teams' synchronization. Submitted (2022)
2. Akjiratikarl, C., Yenradee, P., Drake, P.R.: PSO-based algorithm for home care worker scheduling in the UK. Comput. Ind. Eng. **53**(4), 559–583 (2007)
3. Bean, J.C.: Genetic algorithms and random keys for sequencing and optimization. ORSA J. Comput. **6**(2), 154–160 (1994)
4. Cissé, M., Yalçındağ, S., Kergosien, Y., Şahin, E., Lenté, C., Matta, A.: OR problems related to Home Health Care: a review of relevant routing and scheduling problems. Oper. Res. Health Care **13**, 1–22 (2017)
5. Corbett, J., d'Angelo, C., Gangitano, L., Freeman, J.: Future of health: findings from a survey of stakeholders on the future of health and healthcare in England. Rand Health Quart. **7**(3) (2018)
6. Decerle, J., Grunder, O., El Hassani, A.H., Barakat, O.: A hybrid memetic-ant colony optimization algorithm for the home health care problem with time window, synchronization and working time balancing. Swarm Evol. Comput. **46**, 171–183 (2019)
7. Di Mascolo, M., Martinez, C., Espinouse, M.L.: Routing and scheduling in home health care: a literature survey and bibliometric analysis. Comput. Ind. Eng. **158**, 107255 (2021)
8. Dolan, E., Moré, J.: Benchmarking optimization software with performance profiles. Math. Prog. **91**(2), 201–213 (2002)
9. Du, G., Liang, X., Sun, C.: Scheduling optimization of home health care service considering patients' priorities and time windows. Sustainability **9**(2), 253 (2017)
10. Fikar, C., Hirsch, P.: Home health care routing and scheduling: a review. Comput. Oper. Res. **77**, 86–95 (2017)
11. Gomes, M.I., Ramos, T.: Modelling and (re-)planning periodic home social care services with loyalty and non-loyalty features. Eur. J. Oper. Res. **277**(1), 284–299 (2019)
12. Gonçalves, J., Resende, M.: Biased random-key genetic algorithms for combinatorial optimization. J. Heuristics **17**(5), 487–525 (2011)
13. Gonçalves, J., Resende, M.: Random key genetic algorithms. In: Handbook of Heuristics, Springer, Cham (2018)
14. Grieco, L., Utley, M., Crowe, S.: Operational research applied to decisions in home health care: a systematic literature review. J. Oper. Res. Soc. **72**(9), 1960–1991 (2021)

15. Kummer N, A. F., Buriol, L.S., de Araújo, O.C.: A biased random key genetic algorithm applied to the VRPTW with skill requirements and synchronization constraints. In: Proceedings of the 2020 Genetic and Evolutionary Computation Conference, pp. 717–724 (2020)
16. Liu, R., Tao, Y., Xie, X.: An adaptive large neighborhood search heuristic for the vehicle routing problem with time windows and synchronized visits. Comput. Oper. Res. **101**, 250–262 (2019)
17. Mankowska, D., Meisel, F., Bierwirth, C.: The home health care routing and scheduling problem with interdependent services. Health Care Manag. Sci. **17**(1), 15–30 (2014)
18. Raidl, G.R., Puchinger, J., Blum, C.: Metaheuristic hybrids. In: Handbook of metaheuristics, pp. 385–417. Springer, Cham (2019)
19. Rest, K.D., Hirsch, P.: Daily scheduling of home health care services using time-dependent public transport. Flex. Serv. Manuf. J. **28**(3), 495–525 (2016)
20. Santos, F., Pache, A.C., Birkholz, C.: Making hybrids work: aligning business models and organizational design for social enterprises. Calif. Manag. Rev. **57**(3), 36–58 (2015)
21. Sevaux, M., Sörensen, K., Pillay, N.: Adaptive and multilevel metaheuristics. In: Handbook of Heuristics. Springer, Cham (2018)
22. Shi, Y., Boudouh, T., Grunder, O.: A hybrid genetic algorithm for a home health care routing problem with time window and fuzzy demand. Expert Syst. with Appl. **72**, 160–176 (2017)
23. Wolpert D., Macready W.: No free lunch theorems for optimization. IEEE Trans. Evol. Comput. **1**(1), 67–82 (1997)
24. Spears, W.M., De Jong, K.A.: An analysis of multi-point crossover. Foundations of Genetic Algorithms, vol. 1, pp. 301–315. Elsevier (1991)

An Integer Programming Approach for Sensor Location in a Forest Fire Monitoring System

Beatriz Flamia Azevedo, Filipe Alvelos, Ana Maria A. C. Rocha, Thadeu Brito, José Lima, and Ana I. Pereira

Abstract Forests worldwide have been devastated by fires. Forest fires cause incalculable damage to fauna and flora. In addition, a forest fire can lead to the death of people and financial damage in general, among other problems. To avoid wildfire catastrophes is fundamental to detect fire ignitions in the early stages, which can be achieved by monitoring ignitions through sensors. This work presents an integer programming approach to decide where to locate such sensors to maximize the coverage provided by them, taking into account different types of sensors, fire hazards, and technological and budget constraints. We tested the proposed approach in a real-world forest with around 7500 locations to be covered and about 1500 potential locations for sensors, showing that it allows obtaining optimal solutions in less than 20 min.

B. F. Azevedo (✉) · T. Brito · J. Lima · A. I. Pereira
Research Centre in Digitalization and Intelligent Robotics (CeDRI), Instituto Politécnico de Bragança, Bragança 5300-253, Portugal
e-mail: beatrizflamia@ipb.pt

T. Brito
e-mail: brito@ipb.pt

J. Lima
e-mail: jllima@ipb.pt

A. I. Pereira
e-mail: apereira@ipb.pt

B. F. Azevedo · F. Alvelos · A. M. A. C. Rocha · A. I. Pereira
ALGORITMI Centre, University of Minho, Campus Azurém, Guimares 4800-058, Portugal
e-mail: falvelos@dps.uminho.pt

A. M. A. C. Rocha
e-mail: arocha@dps.uminho.pt

T. Brito
Faculty of Engineering of University of Porto, Porto 4200-465, Portugal

T. Brito · J. Lima
INESC TEC—INESC Technology and Science, Porto 4200-465, Portugal

J. P. Almeida et al. (eds.), *Operational Research*, Springer Proceedings in Mathematics & Statistics 411,
https://doi.org/10.1007/978-3-031-20788-4_2

Keywords Forest fires · Sensors location · Integer programming · Geographic information system (GIS) · Technological and budget limitations

1 Introduction

Forest fires are a severe threat to both natural ecosystems and human beings since forests have an essential role in the global environmental and recreational system, such as atmospheric carbon absorption, soil erosion reduction, moderation of the temperature, and regulation of rainfall [1]. So, perturbation into this ecosystem provokes enormous impacts on fauna and flora, causing economic losses and people's deaths, among other problems.

Globally, it is estimated that humans are responsible for around 75% of all forest fires and much of the increase in fire incidents during 2020 can be directly linked to human actions [17]. With the advent of COVID-19 pandemic, some governments diverted resources to the front line fight against the virus, and the forest patrols and enforcement has been scaled back or stopped altogether [17]. Consequently, the forests were even more unprotected, increasing the number of fire alerts across the globe. From the Amazon to the Arctic, in April 2020, the number of fire alerts across the world was up by 13% compared to last year - which was already a record year for fire [17].

In Europe, the Mediterranean region is the most affected by forest fire catastrophes [15]. Portugal, part of this region, is the country with the highest incidence of wildfires. According to [16], in 2017, Portugal registered 19105 rural fires, resulting in a burnt area of 537143 ha, causing more than 100 human deaths. The following year, 2018, was registered 11450 rural wildfires, with a burnt area of approximately 44078 ha. In 2019, were registered 10841 occurrences of rural fires, corresponding to 41622 ha of the burn area [16]. Although these numbers are decreasing, the number of fires and the burnt area annually recorded are still very high, which generates a high economic, environmental, and humanitarian impact on the affected regions.

There is an urgent need to develop strategies that can serve to map the occurrences and damages caused by forest fires in this context. One possible approach is through the development of forest monitoring systems via remote sensing [6, 7]. This type of system consists of spreading a set of sensors in the forest environment to collect data. The data can be used both to monitor the environment under normal conditions (without fire occurrence) through temperature, humidity, and CO_2 level and quickly detect anomalies, such as fires, through flame or smoke sensors.

Forest monitoring systems provide authorities support in management, planning, resource, location, pre-fire planning, and emergency decision support. In a forest fire situation, a system like this can be determinant since, according to [8, 26], the maximum time interval, from ignition to firefighters' alert response, should not exceed 6 min. Otherwise, the fire will be out of control due to the fire's fast propagation speed.

This paper focuses on defining the location of sensors inside the forest environment to constitute a forest fire monitoring system. For this, an integer programming model was developed, in which some sensors with different technical specifications (cost, range distance) are considered, and the forest characteristic (forest density and forest fire hazard) is taken into account to define the optimum position for each sensor.

Location problems have been addressed by optimization, in particular, by integer programming, since the 1960s. For an overview of the topic, we direct the reader to the comprehensive book Laporte et al. [19]. Although some work has been done on locating fire-related resources (e.g. vehicles [11] and stations and trucks in [21]), decisions related to where to locate sensors in an actual landscape, to the best of our knowledge, have not been addressed before.

The remainder of this paper is organized as follows: after the introduction, Sect. 2 presents the state of art of the mixed-integer programming models in the context of forest fire monitoring systems. After that, the parameters and the region used to build the mathematical model are presented in Sect. 3, and the methodology adopted is described in Sect. 4. Section 5 presents the results and discussion of the approach developed. Section 6 concludes the paper and identifies future paths for further research on this subject.

2 State of Art

Forest fires are considered complex events in causes, intensity, behavior, variability, control, size, and severity. Thus, their early detection is crucial since the time of response will determine the level of damage [9, 24]. For this purpose, at least since the 1960s, as surveyed in [22], research has been done to support decision-making to prevent, detect, monitor, and control forest fires with integer programming.

In [4] a mixed-integer linear program is used to model the spatial fire behavior interacting with suppression placement. The authors defined the study area as cells, and the fire behavior and suppression placement decision are modeled using nodes associated with the cell centers from raster landscapes. The proposed model evaluates fire arrival times and fire lines intensities based on the direction that a fire spreads into a cell as a response to spatially explicit suppression placement. The information of fire spread rates is defined based on Rothermel's equation, and the maximum fire line intensity is based on Byram's fire line intensity [12]. The model presented also considers "control locations areas" that represent the fire suppression resources, modeled as decision variables that alter fire spread paths. Thus if a control area is located at a flammable node, it is assumed that fire will not spread into that node.

A similar approach can be seen in [27]. The authors introduced an algorithm based on the Delaunay triangulation, shortest path algorithms, and mesh refinement to evaluate surface wildfire propagation through a complex heterogeneous landscape. Geographic Information System (GIS) [23] was used to create a spatial model of the region, which includes data about fuel, topography, and weather conditions. To compute the dynamics of fire perimeter extension, a fire perimeter growth construction method to locate the fire perimeter location is utilized. A fire spread model is also used to evaluate the maximum rate of fire spread and other fire parameters. In this approach, each cell's wind and moisture conditions are defined as constant for some time to ensure continuous fire environments for the polygons used in the methods to delimit the area.

A stochastic model of frame spreading prediction is present on [14]. Although wind speed is considered an unpredictable phenomenon, the proposed model can deal with the unpredictable changes in wind speed. The authors considered this problem a stochastic shortest path problem. The landscape is represented as a graph network and the fire propagation time is associated with probabilities for the wildfire arrival time at a point of interest (residences, firemen camp, etc.). To solve the proposed model, the Monte-Carlo simulation is used, and a network size reduction methodology is introduced to optimize the network, removing the redundant edges to speed up the simulation time.

Another interesting work is found in [25], which proposes an integer linear programming model aiming to select the optimal resources to be applied during a planning period for forest fire extinction. In this case, historical data is used to obtain the parameters of cost and resource performance.

In [2] four models integrating fire spread in mixed integer programming are presented in order to solve the location of the optimal resources for the fire forest problem. The first one is for protecting areas, the second for minimizing burned areas, and two others considering fire containment problems.

Resource location in large areas such as forests is hard due to the numerous possibilities to locate them. Thus, developing strategies based on mathematical models to define the optimal position that maximizes forest protection is an important area of study to support fighting the fire combats. Considering the state of the art presented, the mixed-integer linear program is a meaningful approach to deal with the sensor location problem; thus it will be used in this work. Integer programming, since its beginnings in the late 1950s [18] has been used in location analysis [20].

3 Case Study: Experimental Forest Region

The methodology to be developed in this work, will be applied in the region named "Serra da Nogueira", located in the municipality of Bragança, Portugal, as shown in Fig. 1.

The "Serra da Nogueira" is composed of approximately 13 km^2, which represents a large territorial extension and a complex problem for the implementation of a forest monitoring system. For this reason, an experimental region was defined to carry out the methodology presented in this work. Thus, an area of nearly $246,875\,\text{m}^2$ is defined as a forest experimental region, which is presented in Fig. 2. This region was strategically defined due to its heterogeneity, such as different fire hazards, density levels and presence of agriculture or non-agricultural areas.

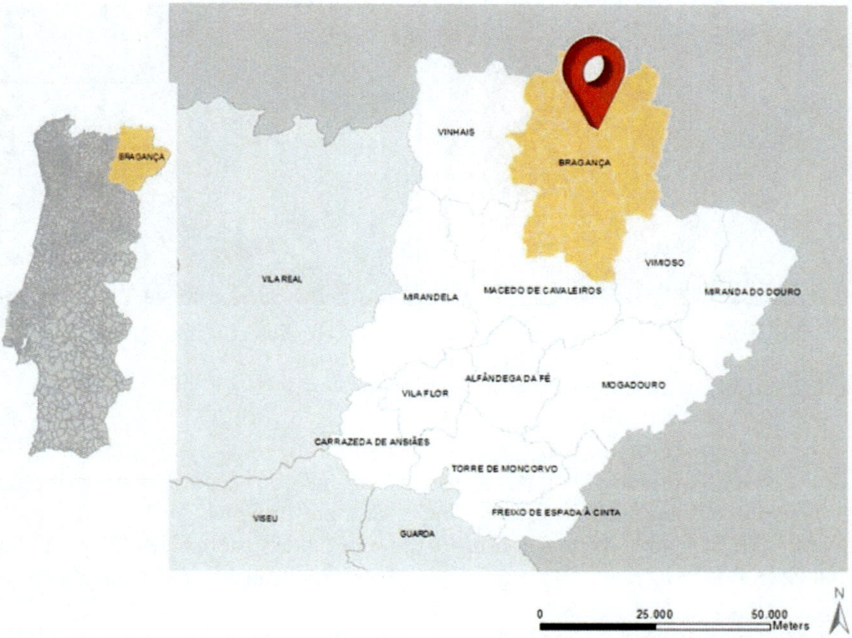

Fig. 1 Serra da Nogueira location

Fig. 2 Forest experimental region

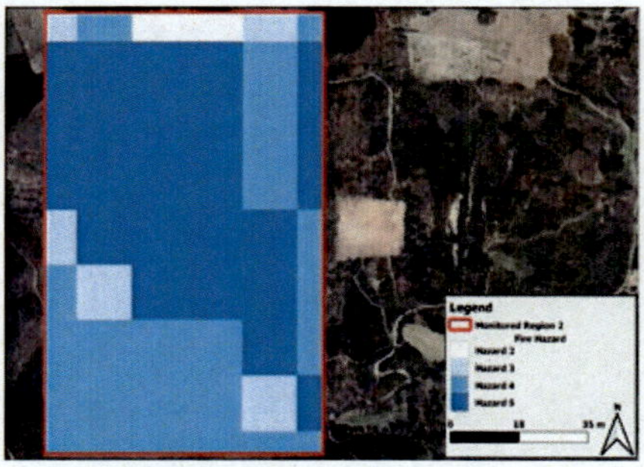

(a) Forest fire hazard map.

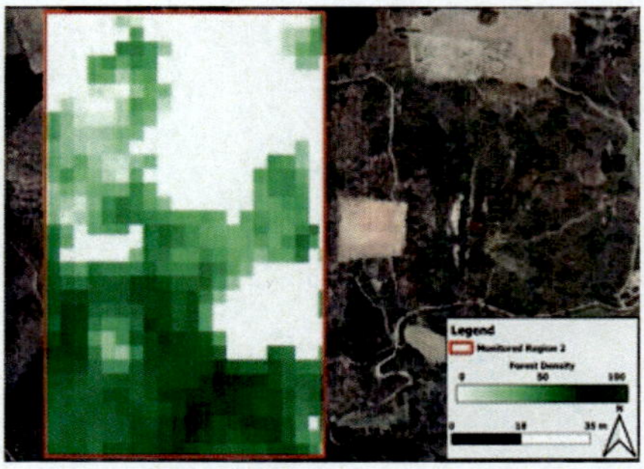

(b) Forest density map.

Fig. 3 Forest experimental regions

By the QGIS software [23], it was possible to map this region, considering the fire hazard and forest density levels. These data are provided by ICNF [16] and Copernicus [10], respectively, according to the coordinate system ETRS89/PT-TM06 (EPSG:3763) UTM Zone 29N standard with Mercator Transverse Universal projection.

The fire hazard can be described as the probability of the event occurrence associated with the terrain conditions. As presented in [28], the fire hazard encompasses two dimensions, space and time, which are intrinsically related to probability and susceptibility. The probability assessment can be based on the historical data of the event for the region that can be considered an uncertain indicator of the fire occurrence [29]. On the other hand, the susceptibility is addressed to aspects of the terrain considered [29]. A territorial unit will be more or less susceptible as affected or potentiates the phenomenon's occurrence and development. In the case of forest fires, a given area will be more susceptible the better it allows the deflagration and/or the progression of the fire spreading [30]. According to [29], it is possible to estimate a fire hazard scale from 0 to 5, according to the previous information presented. In this way, a level 0 indicates a low fire hazard, and level 5 indicates a high hazard fire. Another parameter used is the forest density, which describes the size of the vegetation coverage inside the region demarcated. In this work, the forest density varies from 0 to 100, where 0 indicates no presence of vegetation and 100 indicates a high concentration of vegetation in 40 m^2 can be verified in [3]. Figure 3a shows the forest fire hazard map, whereas Fig. 3b presents the forest density map of the region considered. Note that fire hazard level is not strictly dependent on the forest density since it is a variable related to many other parameters; as previously presented, thence the regions with higher fire hazards are not the same areas with the highest concentration of vegetation.

4 Problem Definition and Mathematical Model

The problem herein described aims to decide where to locate a set of sensors of different types to cover the maximize the coverage (weighted by a hazard index). The region to be monitored was demarcated by cells of 5 m^2, with, consequently, 5 meters of distance between the centers of adjacent cells. This measure was defined by technical experiments, to evaluate the behavior of the sensors over different distances and constraints of the problem [5, 7].

The cell's central point is represented by a node, thus considers, a given sensor k can be placed on a cell j, for $j = 1, ..., n$, with a given coverage that depends on the forest density parameter. When a sensor k, for $k = 1, ..., l$, is assigned to a cell j, it is necessary to identify which cells i are covered by this sensor. It is important to mention that it is considered that each sensor is capable of covering points in any direction, i.e., in 360 degrees. To define if a sensor covers a node, firstly, the Euclidean Distance, d_{ji}, between the cells' nodes c, inside the map, is evaluated by (1).

$$d_{ji} = ||c_j - c_i||_2, \quad j = 1, ..., n; \quad i = 1, ..., n. \tag{1}$$

However, the coverage sensor distance function, $v^k(c_j, c_i)$, is the k sensor view and it depends on the forest density of the cell where the sensor is located f_j^d, the forest density of the cell that can be covered f_i^d, and also the sensor maximum covered distance d_{max}^k. Thereby, the coverage sensor distance is given by Eq. (2).

$$v^k(c_j, c_i) = d_{max}^k \times \left(1 - \frac{f_j^d + f_i^d}{2 f_{max}^d}\right) \tag{2}$$

If the distance (d_{ji}) between the sensor located on c_j and the cell c_i is smaller than the coverage sensor distance $v^k(c_j, c_i)$, the cell c_i is covered by the sensor k placed on c_j cell, that is Eq. (3),

$$d_{ji} \leq v^k(c_j, c_i), \tag{3}$$

In our model, this information is represented by setting the parameter $a_{ij}^k = 1$, if sensor k located in j covers cell i; $a_{ij}^k = 0$ otherwise. Besides this, the following notation is introduced:

- n number of nodes (cells);
- l number of sensors that can be assigned to cells;
- c_k unit cost of sensor k, $k = 1, ..., l$;
- b available budget for sensors;
- h_i hazard of cell i, $i = 1, ..., n$.

Decision variables:

- $y_j^k = 1$, if sensor k ($k = 1, ..., l$) is located in cell j ($j = 1, ..., n$); 0 otherwise;
- $x_i = 1$, if a cell i is covered; 0 otherwise;

Objective function:

$$Max \quad z = \sum_{i=1}^{n} h_i x_i \tag{4}$$

Subject to:

$$\sum_{j=1}^{n} y_j^k \leq 1, \quad k = 1, ..., l \tag{5}$$

$$\sum_{k=1}^{l} y_j^k \leq 1, \quad j = 1, ..., n \tag{6}$$

$$\sum_{k=1}^{l} c_k \sum_{j=1}^{n} y_j^k \leq b \tag{7}$$

$$x_i \leq \sum_{j=1}^{n} \sum_{k=1}^{l} a_{ij}^k y_j^k, \quad i = 1, ..., n \tag{8}$$

$$x_i \in \{0, 1\}, \quad i = 1, ..., n \tag{9}$$

$$y_j^k \in \{0, 1\}, \quad j = 1, ..., n; \quad k = 1, ..., l \tag{10}$$

Objective function (4) maximizes the forest fire hazard covered. Constraints (5) and (6) state that a sensor is not used or is located in a single cell and each cell can accommodate at most one sensor. Constraint (7) is a budget constraint. Constraint (8) is the covering constraint, stating that a covered location must be within the distance of at least one sensor. Equations (9) and (10) are integrability constraints.

5 Results

The sensors will be fixed on the tree trunks, so at least one tree is required on the cell indicated by the solution. In this sense, only points with forest density over or equal to 80 were considered candidates to receive a sensor. This value also ensures appropriate trees in the region to fix the sensors. Moreover, only the cell with a forest fire hazard equal to 5 can receive a sensor in this approach. Thence, after a filter in the cell of the original map, 1499 remains cells on the map to locate the sensors,

Table 1 Available sensor types

Sensor Type	Quantity Available	Unit Cost (€)	Coverage Radio (m)
A	10	45	15
B	5	80	30
C	7	180	100
D	5	350	200
E	3	1000	500

being 7495 the sum of forest fire hazard considering all points available. Five types of sensors were considered, in different quantities, cost, and maximum coverage range, as presented in Table 1.

It is important to mention that the coverage distance varies according to the forest density, as mentioned in Sect. 4. Thence, the values presented in Table 1 correspond to the maximum value that the sensor can reach when there is no forest density interference. In practice, the reached values are defined by Eq. (2).

A maximum budget was stipulated to be spent on purchasing and installing sensors. In this way, a solution defines the optimal number and sensor types according to the budget for each experiment.

Gurobi interface for Python was used to define the model and the general purpose mixed integer programming solver, the Gurobi 9.5 [13], was used for the optimization in an Intel(R) i5(R) CPU @1.60 GHz with 8 GB of RAM machine. Python was used to manipulate the data structures involved.

5.1 Results of Experiment 1

On the first experiment the budget considered was 2000 euros. Figure 4 presents the results of the experimental region, having 5 sensors located: sensor 1—type C, sensor 2—type B, sensor 3 and 4—type D, sensor 5—type E.

The cells (or areas) demarcated by green are regions covered by at least one sensor, while any located sensor does not cover regions in red. It is essential to clarify that the white areas correspond to cells that are not eligible to receive sensors due to the forest density or the forest fire hazard constraints already mentioned.

Through the proposed layout, it can be highlighted that the optimal solution is to locate the more extended range sensors in the areas where there were more points to be covered to maximize the forest fire hazard covered (e.g. sensor 5, which has greater coverage capacity, was positioned in the central region where there are more points to be covered, while sensor 1 and 2, which have less coverage, are in areas with fewer points blue. Another point that can be observed is that although sensor 1 is in a region of a high concentration of points to be covered, the budget constraint

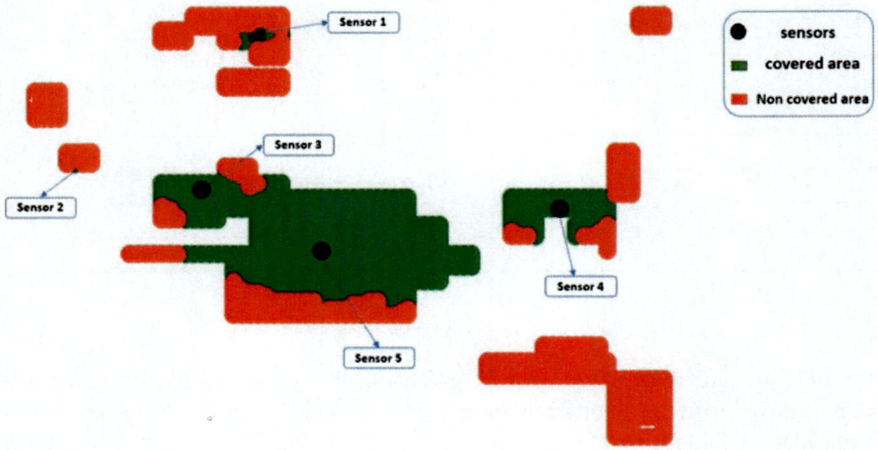

Fig. 4 Results considering a budget equal to 2000 euros

did not support the allocation of a higher value sensor and, consequently, greater coverage capacity.

The cost of this layout is 1960 euros providing coverage of 932 cells, which corresponds to a fire hazard of 4660 units. In this way, it is possible to reduce 62.17 % in the region's total fire hazard. It is important to mention that the exact sensor's location can be found confronting the cell location with the map coordinates, considering the grid pre-established. The optimal solution for the default relative optimality gap of $1e-4$ was obtained in 98 seconds (GPU time), using an 8-core processor and 2316 nodes were explored in 46358 simplex iterations.

5.2 *Results of Experiment 2*

On the second experiment, the budget considered is 4000 euros. By this way 10 sensors was located as presents in Fig. 5: sensor 1—type A, sensor 2—type D, sensor 3, 4 and 5—type C, sensor 6 and 7—type D, sensor 8—type E, sensor 9—type D, sensor 10—type E.

In the second experiment, we have the same region used in the first, but this time we have a higher budget. However, the same behavior observed in the first experiment can be observed in the second one; that is, the longer-range sensors are allocated in regions with the highest concentration of points to maximize the sum of forest fire hazards over surveillance.

The cost of this layout is 3985 euros, providing coverage of 1253 cells, which corresponds to a forest fire hazard of 6280 units. In this way, it is possible to reduce

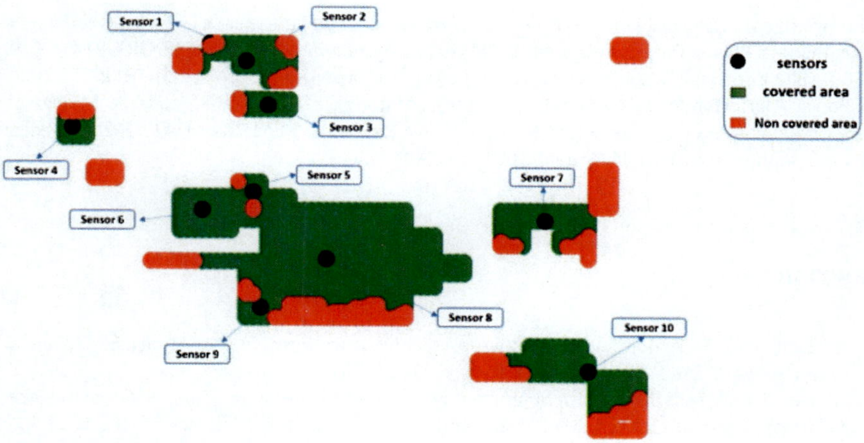

Fig. 5 Results considering a budget equal to 4000 euros

83.79% in the region forest fire hazard. In this experiment, the optimal solution, for the default relative optimality gap of $1e-4$, was obtained in 65 seconds (GPU time), using an 8-core processor and 1373 nodes were explored in 75411 simplex iterations.

6 Conclusion

Forest fire causes environmental disasters and physical and financial damage to the entire ecosystem. For this reason, studying the topic and developing solutions are of great relevance. Currently, multiple techniques and strategies are being searched, proposed, and implemented to solve the emergence problem of wildfires. However, finding an efficient solution to deal with forest fires and replicating them in different regions is not a simple task due to the forest environment's complex dynamics.

This work was conducted to solve the problem of locating wireless sensors in a forest to detect fire ignitions. Being an established approach for location problems, an integer programming model was developed and tested. The results demonstrate that the methodology under development has great potential to assist a decision support system of forest fire detection, in terms of optimal resource location, in this case, sensors.

The proposed integer programming model obtained optimal solutions to a case study in the region of Bragança in less than 20 min, which is acceptable given the time horizon of the decision-making process. The model can be extended to deal with several variants, including (i) using different weights for cells with a different scale from the one used (e.g., increasing the relative importance of low-risk cells) and (ii) minimizing the cost with constraints stating which cells must be covered.

Acknowledgements This work has been supported by FCT Fundação para a Ciência e Tecnologia within the R&D Units Project Scope UIDB/00319/2020 and within project PCIF/GRF/0141/2019 "O3F - An Optimization Framework to reduce Forest Fire" and also the project UIDB/05757/2020 and Forest Alert Monitoring System (SAFe) Project through PROMOVE - Fundação La Caixa. Beatriz Flamia Azevedo is supported by FCT Grant Reference SFRH/BD/07427/2021, Thadeu Brito was supported by FCT PhD grant SFRH/BD/08598/2020.

References

1. Alkhatib, A.A.A.: Article: smart and low cost technique for forest fire detection using wireless sensor network. Int. J. Comput. Appl. **81**(11), 12–18 (2013)
2. Alvelos, F.P.: Mixed integer programming models for fire fighting. In: Computational Science and Its Applications - ICCSA 2018 - 18th International Conference, Melbourne, VIC, Australia, July 2-5, 2018, Proceedings, Part II. Lecture Notes in Computer Science, vol. 10961, pp. 637–652. Springer (2018). https://doi.org/10.1007/978-3-319-95165-2_45
3. Azevedo, B.F., Brito, T., Lima, J., Pereira, A.I.: Optimum sensors allocation for a forest fires monitoring system. Forests **12**(4) (2021). https://doi.org/10.3390/f12040453
4. Belval, E.J., Wei, Y., Bevers, M.: A mixed integer program to model spatial wildfire behavior and suppression placement decisions. Can. J. For. Res. **45**, 384–393 (2015). https://doi.org/10.1139/cjfr-2014-0252
5. Brito, T., Pereira, A.I., Lima, J., Valente, A.: Wireless sensor network for ignitions detection: an IoT approach. Electronics **9**, 1–16 (2020)
6. Brito, T., Zorawski, M., Mendes, J., Azevedo, B.F., Pereira, A.I., Lima, J., Costa, P.: Optimizing data transmission in a wireless sensor network based on lorawan protocol. In: Pereira A.I. et al. (eds) Optimization, Learning Algorithms and Applications. OL2A 2021. Communications in Computer and Information Science. Lecture Notes in Computer Science, vol. 1488, pp. 281–293. Springer (2021). https://doi.org/10.1007/978-3-030-91885-9_20
7. Brito, T., Azevedo, B.F., Valente, A., Pereira, A.I., Lima, J., Costa, P.: Environment monitoring modules with fire detection capability based on IoT methodology. In: Paiva, S., Lopes, S.I., Zitouni, R., Gupta, N., Lopes, S.F., Yonezawa, T. (eds.) Science and Technologies for Smart Cities, pp. 211–227. Springer International Publishing, Cham (2021)
8. Catry, F.X., Moreira, F., Pausas, J.G., Fernandes, P.M., Rego, F., Cardillo, E., Curt, T.: Cork oak vulnerability to fire: The role of bark harvesting, tree characteristics and abiotic factors. PLOS ONE **7**(6), 1–9 (2012)
9. Catry, F.X., Rego, F.C., Bação, F.L., Moreira, F.: Modeling and mapping wildfire ignition risk in Portugal. Int. J. Wildland Fire **18**(8), 921–931 (2010)
10. Copernicus: European union's earth observation programme (2019). https://www.copernicus.eu. Accessed August, 2020
11. Dimopoulou, M., Giannikos, I.: Spatial optimization of resources deployment for forest-fire management. Int. Trans. Oper. Res. **8**, 523–534 (2001)
12. Finney, M.: An overview of flammap fire modeling capabilities, pp. 213–220 (2006). https://www.fs.fed.us/rm/pubs/rmrs-p041/rmrs-p041-213-220.pdf, Accessed on May, 2022
13. Gurobi: Gurobi optimizer (2021). https://www.gurobi.com/products/gurobi-optimizer/. Accessed May, 2021
14. Hajian, M., Melachrinoudis, E., Kubat, P.: Modeling wildfire propagation with the stochastic shortest path: a fast simulation approach. Environ. Modell. Softw. **82**, 73–88 (2016). https://doi.org/10.1016/j.envsoft.2016.03.012
15. Hernández, L.: The Mediterranean burns: WWF's Mediterrenean proposal for the prevention of rural fires. WWF: Gland, Switzerland (2019). http://awsassets.panda.org/downloads/wwf__the_mediterranean_burns_2019_eng_final.pdf. Accessed on May, 2021

16. ICNF: 8 Relatório de Incêncios Rurais 2019 - 01 de Janeiro a 15 de Outubro (2019). http://www2.icnf.pt/portal/florestas/dfci/Resource/doc/rel/2019/2019-10-16-RPIR-08-01jan-15out.pdf Accessed from 21 Oct 2020. . http://www2.icnf.pt/portal/florestas/dfci/Resource/doc/rel/2019/2019-10-16-RPIR-08-01jan-15out.pdf Accessed from 21 Oct 2020.
17. Jeffries, E., Perry, C.: Fires, forest and the future: a crisis raging out of control? (2020). https://wwfeu.awsassets.panda.org/downloads/wwf-fires-forests-and-the-future-report.pdf. Accessed on May, 2021
18. Jünger, M., Liebling, T.M., Naddef, D., Nemhauser, G.L., Pulleyblank, W.R., Reinelt, G., Rinaldi, G., Wolsey, L.A. (eds.): 50 Years of Integer Programming 1958-2008 - From the Early Years to the State-of-the-Art. Springer (2010). https://doi.org/10.1007/978-3-540-68279-0
19. Laporte, G., In Nickel, S., Saldanha, G.F.: Location Science. Springer (2015). https://doi.org/10.1007/978-3-319-13111-5
20. Laporte, G., In Nickel, S., Saldanha, G.F.: Location Science. Springer (2020)
21. Marianov, V., ReVelle, C.: The capacitated standard response fire protection sitting problem: Deterministic and probabilistic models. Ann. Oper. Res. **40**(14), 303322 (1993). https://doi.org/10.1007/BF02060484
22. Martell, D.L.: A review of operational research studies in forest fire management. Can. J. For. Res. **12**(2), 119–140 (1982). https://doi.org/10.1139/x82-020
23. QGIS: A free and open source geographic information system (2019). https://qgis.org. Accessed March, 2021
24. Rego, F., Catry, F., Maia, M., Santos, T., Gravato, A., Castro, I., Moreira, F., Pinto, P., Almeida, J.: Análise da rede nacional de postos de vigia em portugal. Relatório Final do Projecto. ADISA/CEABN-INESC/INOVAÇÃO. Iniciativa Incêndios Florestais, COTEC Portugal (2004)
25. Rodríguez-Veiga, J., Ginzo-Villamayor, M.J., Casas-Mndez, B.: An integer linear programming model to select and temporally allocate resources for fighting forest fires. Forests **9**(10) (2018). https://doi.org/10.3390/f9100583
26. Silva, J.S., Rego, F.C., Fernandes, P., Rigolot, E.: Towards integrated fire management. Outcomes of the European Project Fire Paradox, European Forest Institute (2010)
27. Stepanov, A., Smith, J.M.: Modeling wildfire propagation with delaunay triangulation and shortest path algorithms. Eur. J. Oper. Res. **218**(3), 775–788 (2012). https://doi.org/10.1016/j.ejor.2011.11.03
28. Varnes, D.J.: Landslide hazard zonation: a review of principles and practices. UNESCO - Paris (1984)
29. Verde, J.C.: Avaliação da perigosidade de incêndio florestal. Ph.D. thesis, University of Lisbon (2010)
30. Zêzere, J.L., Reis, E., Garcia, R., Oliveira, S., Rodrigues, M.L., Vieria, G., Ferreira, A.B.: Integration of spatial and temporal data for the definition of different landslide hazard scenarios in the area north of lisbon (portugal). Nat. Hazard. **4**(1), 133–146 (2004). https://doi.org/10.5194/nhess-4-133-2004

Capacity Allocation Incorporating Market Equity Concerns: A Pharmaceutical Supply Chain Case Study

Catarina Bessa, Raquel Duque, Alexandre Jesus, Cristóvão Silva, Lukas Eberle, and Samuel Moniz

Abstract This work seeks to improve the efficiency of global pharmaceutical supply chains by proposing a capacity allocation model that includes an unfairness metric to balance the response to market needs with economic goals. A three-tier supply chain deterministic model is developed to include both the Net Present Value (NPV) and an unfairness metric that accounts for drug shortages. To ensure the validity of the proposed approach, the model was verified with data collected from a leading pharmaceutical company. Furthermore, results show the impact of equity concerns in product allocation, inventory, and investment in capacity decisions within the supply chain. The trade-off analysis between unfairness and NPV provides an important decision-support tool for evaluating different scenarios. The NPV of a fairer supply chain is significant, which proves the importance and feasibility of the proposed model.

Keywords Capacity allocation · Mathematical programming · Supply chain optimisation · Fairness metrics

1 Introduction

The performance of pharmaceutical supply chains (PSC) directly impacts the healthcare structure and quality of life of the populations. Such impact comprises not only the price at which people have access to medication and healthcare services, but also the continuous availability of those [1]. The latter consequence is often related to the

C. Bessa (✉) · R. Duque · A. Jesus · C. Silva · S. Moniz
Department of Mechanical Engineering, University of Coimbra, Coimbra, Portugal
e-mail: ana.bessa@dem.uc.pt

S. Moniz
e-mail: samuel.moniz@dem.uc.pt

L. Eberle
F. Hoffman-La Roche Ltd., Basel, Switzerland

J. P. Almeida et al. (eds.), *Operational Research*, Springer Proceedings in Mathematics & Statistics 411,
https://doi.org/10.1007/978-3-031-20788-4_3

capability of the supply chain (SC) to deliver drugs, which is the problem addressed in this work.

Many supply interruptions are responsible for drug shortages. Take for example the increase of drug shortages in the United States, from 154 to 456, between 2007 and 2012 [2]. Furthermore, very recent data about covid-19 vaccines show that these have been distributed unequally. Approximately 85% of all vaccines were administered in high and upper middle income countries, and 75% of those vaccines were received by just ten countries [3]. On the one hand, if the same amount of vaccines were given to every country, equality would be applied, but the outcome would be extremely unfair since countries have different population sizes. On the other hand, a fair outcome would be to deliver the same relative amount of vaccines to each country. Another example of equity is the case of food assistance. If the same amount of food is given to every different region, the outcome is highly unfair since some areas will not be able to provide enough food, and others will generate waste. So, fairness is the equity applied to the outcome aimed to create (i.e., provide the same portion of support according to the demand of each area).

In short, including fairness in capacity planning optimisation is critical to ensure that the same value is generated for all involved parts (or stakeholders). Although fairness is a defiant concept from an operations management perspective, it has been described as the feeling of equity among a set of stakeholders [4]. To address the challenge of appraising such a feeling, Fehr and Schmidt [4] measured fairness by comparing the outputs that each stakeholder gets from a specific decision. Those outputs are presented as utility values and computed through utility functions.

While the most common performance indicators in SC optimisation are related to economic goals, our work proposes a bi-objective capacity allocation model that includes an unfairness metric to balance the demand response to the market needs with an economic goal. The proposed model represents a 3-tier PSC consisting of two sequential production stages and markets. Unfairness is considered through a utility function, which measures the degree of inequity in the supply of drugs, as well as the overall shortage. Aiming at minimising unfairness and maximising the economic value generated, trade-offs are obtained in the form of a Pareto front. As a result, the presented decision support tool gives valuable insights into how to allocate capacity in PSCs to reach the desired values of fairness and efficiency.

Managing PSCs such that capacity planning decisions are made considering supply unfairness alongside cost-efficiency can be seen as one of the key PSC challenges. Even so, to the best of our knowledge, supply unfairness has never been addressed in the context of capacity planning of PSC, and equity and cost-efficiency objectives have never been jointly studied in the literature. In the following sections, we present the theoretical background, and we introduce the problem description and model formulation. We then present the results, and the paper concludes with some remarks and insights on future work.

2 Related Literature

2.1 Fairness and Utility Functions

The most widely used method to measure fairness is the unfairness aversion model introduced by Fehr and Schmidt [4], which is quantified by the utility function (1). The authors state that unfairness is strongly correlated to the feeling of inequity among a set of n stakeholders. The utility function sums the total amount of utility (absolute amount of gaining or losing) with the amount of inequity created among the stakeholders. If we consider an example of a monetary payoff, the total amount of utility would be the payoff (x_i) of each stakeholder (i), and the inequity would be the difference between the payoffs. Though, inequity has two different scenarios: one of disadvantage, the second term in (1), and another of advantage, the third term in (1). The advantageous situation concerns the stakeholder who earns more in comparison to the others, while the disadvantageous case concerns the stakeholders who earn less. Both types of inequities are balanced by a sensitivity coefficient, β_i and α_i, respectively. These coefficients might be changed upon the decision-maker perspective of the situation, as analysed by Tao et al. [5]. Overall, these coefficients weight the feeling of unfairness created in the stakeholder when facing a situation of inequity.

The coefficient of advantage (β_i) is usually defined within the interval $0 \leq \beta_i < 1$. Note that β_i is less than 1 to avoid the feeling of superiority (i.e., cases in which the stakeholder would appreciate being in advantage) [4]. Considering a standard outcome, negative deviations count more than the positive ones, thus $\alpha_i \geq \beta_i$. The upper bound on α_i is not needed since the stakeholders in disadvantage are often willing to reduce their monetary payoff if other stakeholders' payoff is decreased even more [4].

Utility function (1) has been widely used in different optimisation problems such as SC facility location, vehicle routing, and project allocation problems [6]. In particular, it has been applied to study the number and budget of projects allocated to a department [7] or to measure the amount of shortage in procurement and distribution decisions [5]. Our approach to model unfairness is based on the work presented by Tao et al. [5].

$$U_i(x) = x_i - \alpha_i \frac{1}{n-1} \sum_{j:i \neq j} max\{x_j - x_i, 0\} - \beta_i \frac{1}{n-1} \sum_{j:i \neq j} max\{x_i - x_j, 0\} \quad (1)$$

2.2 Multi-objective Optimisation Trade-Off Methods

Multi-objective optimisation methods look for a trade-off between the solutions of all the objective functions or Pareto optimal solutions. A Pareto-optimal solution only exists if there is no feasible solution that can improve one objective function without degrading the others. Naturally, this definition allows for the existence of multiple optimal solutions which, when grouped and displayed, form a Pareto front. The methods can be classified as a priori, interactive and a posteriori methods [8]. In a priori methods, the decision-maker sets a goal or weight to the objectives before the resolution. Contrarily, interactive methods are conducted repeatedly, alternating between calculations and dialogues with the decision-maker until convergence. The a posteriori methods only include the decision-maker in evaluating the efficient solutions generated, not discarding possible solutions during iterations, which increases the confidence level of the decision-maker. Despite the computational effort needed, this advantage is the main reason for their popularity in SC optimisation [9].

One of the most common a posteriori methods is the ε-constraint which consists of optimising one of the objective functions, constrained by the others. In this field, Mavrotas [8] proposed the augmented ε-constraint method, AUGMECON, which avoids weakly Pareto-optimal solutions and accelerates the process of convergence. This method has been used to conduct the trade-off analysis between unfairness and cost-efficiency.

3 Problem Description

A 3-tier PSC is considered, as depicted in Fig. 1, where our goal is to optimise the capacity allocation according to the existing trade-off between Net Present Value (NPV) and utility of shortage. To fairly supply the markets, we used the utility function proposed by Tao et al. [5] to consider the utility of shortage, i.e., to account for the amount of unmet drug demand and the discrepancy of demand response among markets.

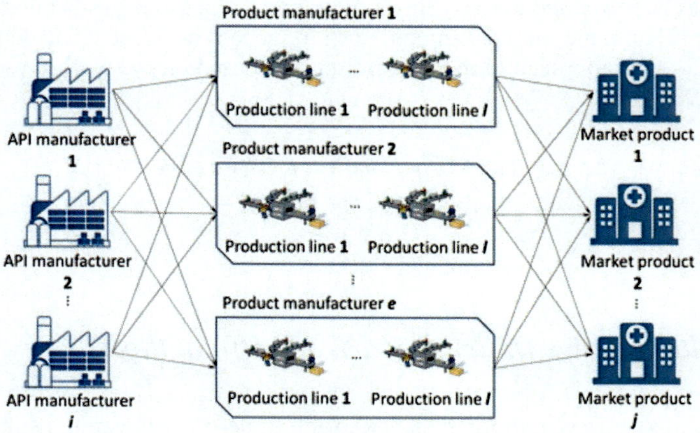

Fig. 1 Structure of the Pharmaceutical Supply Chain

Pharmaceutical production plants typically use a batch and multi-purpose setting to process different products [10]. This is the case of Active Pharmaceutical Ingredient (API) production, while drug product production is usually performed on production lines. For the purpose of this work, it is assumed that the demand for each different product is the demand of a given market. Furthermore, inventory decisions and investment in extra capacity are possible to occur in all stages of the SC.

A production-technology fit is given to describe the ability of a manufacturer or production line to process a specific product. Finally, a simplified product recipe is considered to represent the association between API and product. The drug substance production stage is limited by the API supply and its own capacity.

The main goal is to find the most profitable operation of the network for each value of unfairness. The model suggests decisions concerning the quantity of API and drug products produced at each SC tier, as well as the inventory holding levels and the investments in capacity made at each time period.

3.1 Mathematical Formulation

The indices, sets, parameters and decision variables are defined in Tables 1, 2, 3, 4 and 5. We assume that parameters and variables related to production and capacity are measured in terms of time, e.g., x_{ait} denotes the time spent producing API a in manufacturer i during period t.

Table 1 Indices

i	API manufacturer
e	Product manufacturer
j	Drug product
a	API
t	Time period
l	Production line

Table 2 Sets

$\boldsymbol{I}$	API manufacturers
$\boldsymbol{E}$	Product manufacturers
$\boldsymbol{J}$	Drug products
$\boldsymbol{A}$	APIs
$\boldsymbol{T}$	Time periods
$\boldsymbol{L}$	Production lines

Table 3 Parameters

d_t	Discount factor at period t
R_j	Sales price of product j (monetary units)
C_a^{prod}	Production cost of API a (monetary units)
C_j^{prod}	Production cost of product j (monetary units)
C_a^{inv}	Inventory holding cost of API a (monetary units)
C_j^{inv}	Inventory holding cost of product j (monetary units)
C^{setup}	Setup cost (monetary units)
C_{it}^{new}	Cost of opening capacity at API manufacturer i at period t (monetary units)
C_{elt}^{new}	Cost of opening capacity at product manufacturer e, line l at period t (monetary units)
α^{dis}	Coefficient of disadvantage
β^{adv}	Coefficient of advantage
D_{jt}	Demand of product j during period t (units of capacity)
S^{setup}	API setup time (units of capacity)
U_i^{min}	Minimum capacity utilisation for API manufacturer i (units of capacity)
U_i^{max}	Maximum capacity utilisation for API manufacturer i (units of capacity)
U_e^{min}	Minimum capacity utilisation for product manufacturer e (units of capacity)
U_e^{max}	Maximum capacity utilisation for product manufacturer e (units of capacity)
I_j^{init}	Quantity of product j in inventory at the first period t (units of capacity)
I_a^{init}	Quantity of API a in inventory at the first period t (units of capacity)
θ_{aj}	Required amount of API a for producing one unit of product j
Q_a^{min}	Minimum batch size for API a (units of capacity)
Q_j^{min}	Minimum production for product j (units of capacity)
B_i	Minimum investment in API Manufacturer i (units of capacity)
B_{el}	Minimum investment in product Manufacturer e, production line l (units of capacity)
r	Interest rate

Table 4 Binary variables

s_{ait}	$\begin{cases} 1, \text{ if API } a \text{ is being setup in API manufacturer } i \text{ at period } t, \\ 0, \text{ otherwise} \end{cases}$
y_{it}	$\begin{cases} 1, \text{ if API manufacturer } i \text{ is opened at period } t, \\ 0, \text{ otherwise} \end{cases}$
z_{elt}	$\begin{cases} 1, \text{ if product manufacturer } e, \text{ line } l \text{ is opened at period } t, \\ 0, \text{ otherwise} \end{cases}$
h_{it}	$\begin{cases} 1, \text{ if investment is made in API manufacturer } i \text{ at period } t, \\ 0, \text{ otherwise} \end{cases}$
h_{elt}	$\begin{cases} 1, \text{ if investment is made in product manufacturer } e, \text{ production line } l, \text{ at period } t, \\ 0, \text{ otherwise} \end{cases}$

Table 5 Continuous variables

$m_{jj't}$	Inequity between drugs j and j' at period t (auxiliary variable for linearisation purposes)
η_{jt}	Percentage of unmet demand of product j at period t
w_{jt}	Quantity of product j delivered to market at period t (units of capacity)
x_{ait}	Quantity of API a produced in API manufacturer i at period t (units of capacity)
v_{jelt}	Quantity of product j produced in product manufacturer e, production line l during period t (units of capacity)
k_{it}^{new}	Extra capacity available in API manufacturer i at period t (units of capacity)
k_{elt}^{new}	Extra capacity available in product manufacturer e, production line l at period t (units of capacity)
l_{it}^{new}	Amount of capacity invested in API manufacturer i at period t (units of capacity)
l_{elt}^{new}	Amount of capacity invested in product manufacturer e, line l at period t (units of capacity)
I_{at}	Quantity of API a in inventory at period t (units of capacity)
I_{jt}	Quantity of product j in inventory at period t (units of capacity)

The objective function (2) seeks to maximise the NPV of the entire SC by subtracting the operational costs from the revenue obtained from the drugs delivered to the markets. The cost terms correspond to the production, inventory, setup and installing new capacity, respectively. Each cost is specified for both API and product manufacturers, except for setup costs. While setup time is discriminated for API production, setup costs are considered within the production costs of the drug products.

The objective function (3) minimises the unfairness through the utility of shortage. As referred above, expression (3) considers the shortage amount (first term) and inequity of unmet demand both in the disadvantageous (second term) and advantageous (third term) situations. Note that the disadvantageous situation considers the cases where a specific market has more unmet demand than the others, and the advantageous situation reflects the opposite concern.

$$\begin{aligned} \max \ & \textstyle\sum_{j\in J}\sum_{t\in T} d_t\, R_j\, w_{jt} - \Big(\sum_{a\in A}\sum_{i\in I}\sum_{t\in T} d_t\, C_a^{prod}\, x_{ait} \\ & + \textstyle\sum_{j\in J}\sum_{e\in E}\sum_{l\in L}\sum_{t\in T} d_t\, C_j^{prod}\, v_{jelt} + \sum_{a\in A}\sum_{t\in T} d_t\, C_a^{inv}\, I_{at} \\ & + \textstyle\sum_{j\in J}\sum_{t\in T} d_t\, C_j^{inv}\, I_{jt} + \sum_{a\in A}\sum_{i\in I}\sum_{t\in T} d_t\, C^{setup}\, s_{ait} \\ & + \textstyle\sum_{i\in I}\sum_{t\in T} d_t\, C_{it}^{new}\, l_{it}^{new} + \sum_{e\in E}\sum_{l\in L}\sum_{t\in T} d_t\, C_{elt}^{new}\, l_{elt}^{new}\Big) \end{aligned} \tag{2}$$

$$\begin{aligned} \min \ & \textstyle\sum_{j\in J}\sum_{t\in T} \eta_{jt} + \alpha^{dis}\frac{1}{|J|-1}\sum_{j,j'\in J: j\neq j'}\sum_{t\in T} max(\eta_{jt} - \eta_{j't}, 0) \\ & + \beta^{adv}\frac{1}{|J|-1}\textstyle\sum_{j,j'\in J: j\neq j'}\sum_{t\in T} max(\eta_{j't} - \eta_{jt}, 0) \end{aligned} \tag{3}$$

In order to overcome the non-linearity of the objective function (3), a linearisation was conducted in which the second and third terms were modelled using auxiliary decision variables $m_{jj't}$. In this way, the objective function (3) can be replaced by (4) subject to constraints (5)–(8).

$$\begin{aligned} \min \ & \textstyle\sum_{j\in J}\sum_{t\in T} \eta_{jt} + \alpha^{dis}\frac{1}{|J|-1}\sum_{j,j'\in J: j\neq j'}\sum_{t\in T} m_{jj't} \\ & + \beta^{adv}\frac{1}{|J|-1}\sum_{j,j'\in J: j\neq j'}\sum_{t\in T} m_{j'jt} \end{aligned} \tag{4}$$

$$m_{jj't} \geq \eta_{jt} - \eta_{j't}, \ \forall j, \ j' \in J : j \neq j', \ t \in T \tag{5}$$

$$m_{jj't} \geq 0, \ \forall j, \ j' \in J : j \neq j', \ t \in T \tag{6}$$

$$m_{j'jt} \geq \eta_{j't} - \eta_{jt}, \ \forall j, \ j' \in J : j \neq j', \ t \in T \tag{7}$$

$$m_{j'jt} \geq 0, \ \forall j, \ j' \in J : j \neq j', \ t \in T \tag{8}$$

Constraint (9) assures the balance between drugs delivery and shortage. Capacity constraints are found in (10)–(13). Therefore, the capacity utilisation is held within the desired limits, both in the API and drug product manufacturers. In restrictions

(14) and (15), the available extra capacity is given according to investments made in previous periods of time. Both (16) and (17) represent the inventory balance equalities for API and product manufacturers. Those consist of equalities between the production, product delivered, and prior inventory unused. Equations (18) and (19) set the minimum quantity and batch sizes to produce each API and drug product. Constraints (20) and (22) guarantee that the capacity investments are only made in active facilities, while (21) and (23) are both the activation constraints on the binary variables responsible for the capacity investment as well as the minimum capacity investment constraints. Expression (24) sets the value of the discount factor according to the interest rate of each period. From constraint (25) further, variables are set between the feasible values.

$$\frac{w_{jt}}{D_{jt}} + \eta_{jt} = 1\,,\ \forall j \in \boldsymbol{J},\ t \in \boldsymbol{T} \tag{9}$$

$$\sum\nolimits_{a\in \boldsymbol{A}} \left(x_{ait} + s_{ait}\, S^{setup}\right) \geq U_i^{min}\, y_{it}\,,\ \forall i \in \boldsymbol{I},\ t \in \boldsymbol{T} \tag{10}$$

$$\sum\nolimits_{a\in \boldsymbol{A}} \left(x_{ait} + s_{ait}\, S^{setup}\right) \leq U_i^{max}\, y_{it} + k_{it}^{new}\,,\ \forall i \in \boldsymbol{I},\ t \in \boldsymbol{T} \tag{11}$$

$$\sum\nolimits_{j\in \boldsymbol{J}} v_{jelt} \geq U_e^{min}\, z_{elt}\,,\ \forall e \in \boldsymbol{E},\ l \in \boldsymbol{L},\ t \in \boldsymbol{T} \tag{12}$$

$$\sum\nolimits_{j\in \boldsymbol{J}} v_{jelt} \leq U_e^{max}\, z_{elt} + k_{elt}^{new}\,,\ \forall e \in \boldsymbol{E},\ l \in \boldsymbol{L},\ t \in \boldsymbol{T} \tag{13}$$

$$k_{it}^{new} = \sum\nolimits_{t'\in \boldsymbol{T}:t'<t} l_{it'}^{new}\,,\ \forall i \in \boldsymbol{I},\ t \in \boldsymbol{T} : t > 1 \tag{14}$$

$$k_{elt}^{new} = \sum\nolimits_{t'\in \boldsymbol{T}:t'<t} l_{elt'}^{new}\,,\ \forall e \in \boldsymbol{E},\ l \in \boldsymbol{L},\ t \in \boldsymbol{T} : t > 1 \tag{15}$$

$$I_{at} = \left(I_a^{init}|_{(t=1)},\ I_{a,t-1}|_{(t>1)}\ \right) + \sum\nolimits_{i\in \boldsymbol{I}} x_{ait} - \sum\nolimits_{j\in \boldsymbol{J}} \sum\nolimits_{e\in \boldsymbol{E}} \sum\nolimits_{l\in \boldsymbol{L}} v_{jelt}\, \theta_{aj}\,,\quad \forall a \in \boldsymbol{A},\ t \in \boldsymbol{T} \tag{16}$$

$$I_{jt} = \left(I_j^{init}|_{(t=1)}, I_{j,t-1}|_{(t>1)}\right) + \sum\nolimits_{e\in \boldsymbol{E}} \sum\nolimits_{l\in \boldsymbol{L}} v_{jelt} - w_{jt}\,,\ \forall j \in \boldsymbol{J},\ t \in \boldsymbol{T} \tag{17}$$

$$x_{ait} \geq Q_a^{min}\, y_{it}\,,\ \forall a \in \boldsymbol{A},\ i \in \boldsymbol{I},\ t \in \boldsymbol{T} \tag{18}$$

$$v_{jelt} \geq Q_j^{min}\, z_{elt}\,,\ \forall j \in \boldsymbol{J},\ e \in \boldsymbol{E},\ l \in \boldsymbol{L},\ t \in \boldsymbol{T} \tag{19}$$

$$l_{it-1}^{new} \leq M\, y_{it}\,,\ \forall i \in \boldsymbol{I},\ t \in \boldsymbol{T} : t > 1 \tag{20}$$

$$h_{it}\, M \geq l_{it}^{new} \geq B_i\, h_{it}\,,\ \forall i \in \boldsymbol{I},\ t \in \boldsymbol{T} \tag{21}$$

$$l_{elt-1}^{new} \leq M\, z_{elt}\,,\ \forall e \in \boldsymbol{E},\ l \in \boldsymbol{L},\ t \in \boldsymbol{T} : t > 1 \tag{22}$$

$$h_{elt}\, M \geq l_{elt}^{new} \geq B_{el}\, h_{elt}\,, \forall e \in \boldsymbol{E},\ l \in \boldsymbol{L}, t \in \boldsymbol{T} \tag{23}$$

$$d_t = \frac{1}{(1+r)^t}\,,\ \forall t \in \boldsymbol{T} \tag{24}$$

$$z_{elt}\,,\ h_{elt} \in \{0, 1\}\,\forall e \in \boldsymbol{E}\,,\ l \in \boldsymbol{L}\,,\ t \in \boldsymbol{T} \tag{25}$$

$$s_{ait} \in \{0, 1\}\,,\ x_{ait} \geq 0\,,\ \forall a \in \boldsymbol{A}\,,\ i \in \boldsymbol{I}\,,\ t \in \boldsymbol{T} \tag{26}$$

$$v_{jelt} \geq 0\,,\ \forall j \in \boldsymbol{J}\,,\ e \in \boldsymbol{E}\,,\ l \in \boldsymbol{L}\,,\ t \in \boldsymbol{T} \tag{27}$$

$$l_{elt}^{new}\,,\ k_{elt}^{new} \geq 0\,,\ \forall e \in \boldsymbol{E}\,,\ l \in \boldsymbol{L}\,,\ t \in \boldsymbol{T} \tag{28}$$

$$\eta_{jt} \in [0, 1], \forall j \in \boldsymbol{J}, t \in \boldsymbol{T} \tag{29}$$

$$y_{it}, h_{it} \in \{0, 1\}\forall i \in \boldsymbol{I}, t \in \boldsymbol{T} \tag{30}$$

$$w_{jt}, I_{jt} \geq 0\,\forall j \in \boldsymbol{J}, t \in \boldsymbol{T} \tag{31}$$

$$I_{at} \geq 0, \forall a \in \boldsymbol{A}, t \in \boldsymbol{T} \tag{32}$$

$$l_{it}^{new}, k_{it}^{new} \geq 0, \forall i \in \boldsymbol{I}, t \in \boldsymbol{T} \tag{33}$$

4 Experimental Results and Sensitivity Analysis

We next validate the proposed mathematical model and demonstrate its applicability to the problem. The model was solved using CPLEX 20.1.0. in a computer with Intel® Core™ CPU i7-8550U @ 1.8 GHz processor and 16 GB RAM.

In our case, we considered a PSC producing 3 APIs in 2 different manufacturers that supply 4 drug product manufacturers, each composed of 3 production lines. A total of 6 different products need to be provided over 9 years of operation. Note that not all sites are able to produce all references. The sensitivity coefficients, α^{dis} and β^{adv} were defined to 0,8 and 0,5 respectively, according to the principle stated by Fehr and Schmidt [4] that the stakeholder in disadvantage has a stronger feeling of inequity than the one in advantage. Data collection has been performed to obtain values of the production capacity and efficiency, nonetheless, those cannot be disclosed due to confidentiality reasons.

4.1 Computational Results

This section presents the computational results obtained during the generation of the Pareto front. The method used, AUGMECON, requires the development of two payoff tables, which set the interval in which the non-dominated solutions are found. The first payoff table, Table 6a, sets the individual optima of the objective functions, i.e., objective values when individually optimised. The second payoff table, Table 6b, provides the range of values in which the second objective function is optimal and

Table 6 Pay-off table obtained by **(a)** individual and **(b)** lexicographic optimisation

(a)

	NPV (2)	Unfairness (4)
Max (2)	16.53	46.66
Min (4)	-4.06	4.16

(b)

	NPV (2)	Unfairness (4)
Max (2)	16.53	35.25
Min (4)	-4.06	4.16

guarantees the optimal value of the first objective function. This interval was then divided into ten equally distanced values of the second objective function. Finally, the Pareto front was generated by optimising the first objective constrained to each of these ten values.

Concerning the model performance, numerical results demonstrate that the CPLEX version 20.1.0 could prove optimality for all runs, having a computational time of less than 1 s. Furthermore, when including expressions (4)–(8) in the formulation, the model results in 2480 constraints, 288 binary variables and 1270 continuous variables.

4.2 *Numerical Results and Practical Implications*

The numerical results and practical implications of the conducted optimisation are further analysed in this section. To that end, the NPV resulting from objective function (2) is analysed against unfairness, measured by the utility of shortage given by expression (4). Emphasis should be given to the fact that the utility of shortage is worse when it increases, thus reflecting the feeling of unfairness.

Considering that the 6 products analysed are arranged in increasing order of their profit, e.g., product 1 is less profitable than product 2, and so on, Fig. 2 presents the utility of shortage of each drug product against the overall unfairness created. Figure 2 clearly shows that for high values of unfairness, the model results are focused on maximising NPV, as the most profitable products are chosen to meet the demand independently of the unfairness created. To achieve the minimum utility, all products must attain similar unfairness values.

Figure 3 illustrates the extent of the trade-off between NPV and unfairness, in which higher unfairness values allow for superior profit. This means that, in the most profitable scenarios, a lot of demand is unmet and there is a great discrepancy among different drugs, making it relatively inexpensive to decrease unfairness. This is proven by the highlighted part of the graph, between 13.2 and 16.5 r.m.u. (relative monetary units), where a 43% unfairness reduction is possible with just a 20% NPV loss. After this threshold, potential improvements in unfairness are significantly more costly because all unfairness terms need to be improved.

Figure 4 depicts the installed capacity and utilisation along with the different values of unfairness. Since there is insufficient capacity to meet all the demand, investments in capacity are required to decrease unfairness. It can be seen that the

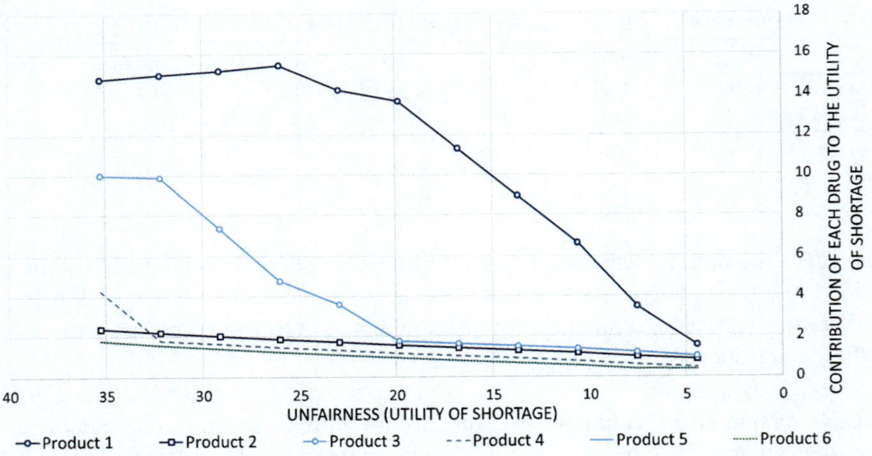

Fig. 2 Utility of shortage per product for each utility of shortage

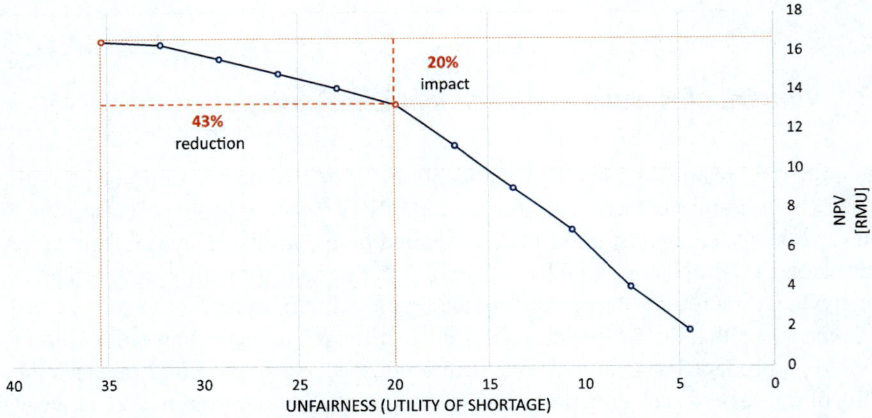

Fig. 3 Pareto-front: The trade-off between NPV and utility of shortage

capacity utilisation increases subtly until up to a utility of 26, and abruptly from this point until 20. From this value, the capacity utilisation is maximised, and the unfairness reduction is achieved at the expense of decreasing the NPV significantly. In other words, to fully incorporate equity concerns in PSC, significant investments in capacity might be required.

Finally, we also show the relationship between unfairness and productivity (Fig. 5). Productivity is a measure of practical interest since it provides the ratio between the value created (outputs) and cost (inputs). As can be seen, decreasing unfairness has a negative impact on productivity. For lower utility values, the same decrease of unfairness can be done with little impact on productivity.

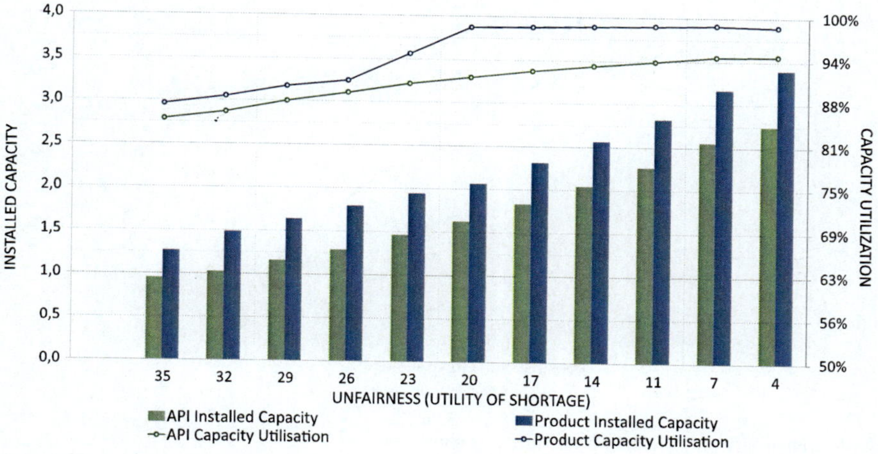

Fig. 4 Capacity utilisation of manufacturers

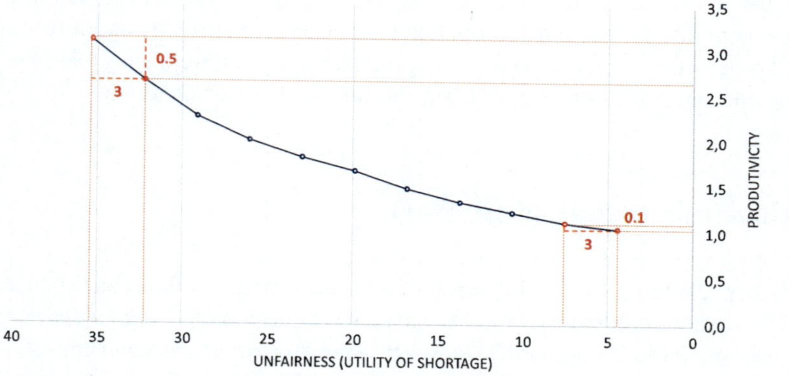

Fig. 5 Productivity according to the utility of shortage

4.3 *Sensitivity Analysis*

To further understand the impact of modelling fairness in the decision-making process, we performed a sensitivity analysis on the utility function (4) as follows:

Scenario I—full equity: both the advantageous and disadvantageous situations are thought to create a feeling of unfairness. The sensitivity coefficients are set as $\alpha^{dis} = 0,8$ and $\beta^{adv} = 0,5$.

Scenario II—equity on disadvantage: the inequity only creates a feeling of unfairness in disadvantageous situations, then, $\alpha^{dis} = 0,8$ and $\beta^{adv} = 0$.

Scenario III—inequity: the weight of the inequity is irrelevant to the utility created. Then, utility is entirely created by meeting the demand. In this scenario, $\alpha^{dis} = 0$ and $\beta^{adv} = 0$.

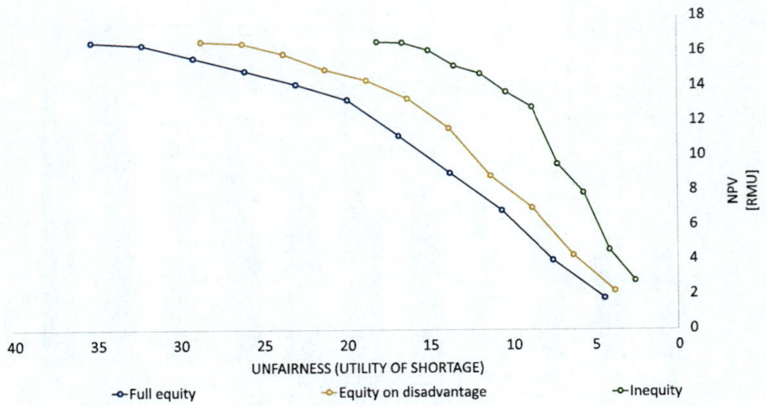

Fig. 6 Sensitivity coefficients on the utility of shortage

Figure 6 presents the Pareto front for these scenarios. The differences between using a common unmet demand objective function (scenario III) and a full utility function objective (scenario I) can be seen. The tinier difference between the first and second scenarios proves that, when focusing on decreasing the

5 Conclusions and Future Work

In this work, we propose the inclusion of a fairness metric in the capacity planning of PSC to tackle supply inequities. To that end, we developed a bi-objective capacity allocation model for 3 stages PSC, aiming to generate cost-efficient and fair solutions. Our results suggest that a significant amount of unfairness can be tackled with little impact on economic targets. Future research should extend the model to include product reallocation decisions within the SC to better follow a fairness strategy without penalising the economic value generated.

Funding

This research is sponsored by national funds through FCT – Fundação para a Ciência e a Tecnologia –, under the projects PTDC/EME-SIS/6019/2020 and UID/EMS/00285/2020.

References

1. Chen, X., Li, S., Wang, X.: International Journal of Production Economics **230**, 107770 (2020). https://doi.org/10.1016/j.ijpe.2020.107770
2. Lücker, F., Seifert, R.W.: Omega **73**, 114 (2017). https://doi.org/10.1016/j.omega.2017.01.001
3. Asundi, A., O'Leary, C., Bhadelia, N.: Cell Host & Microbe **29**(7), 1036 (2021). https://doi.org/10.1016/j.chom.2021.06.007
4. Fehr, E., Schmidt, K.M.: Quart. J. Econ. **114**(3), 817 (1999). https://doi.org/10.1162/003355399556151
5. Tao, J., Shao, L., Guan, Z., Ho, W., Talluri, S.: Int. J. Prod. Res. **58**(7), 1950 (2020). https://doi.org/10.1080/00207543.2019.1637955
6. Park, C.H., Berenguer, G.: Production and Operations Management **29**(11), 2461 (2020). https://onlinelibrary.wiley.com/doi/abs/10.1111/poms.13235
7. Naldi, M., Nicosia, G., Pacifici, A., Pferschy, U.: Soc.-Econ. Plann. Sci. **67**, 133 (2019). https://doi.org/10.1016/j.seps.2018.10.007
8. Mavrotas, G.: Appl. Math. Comput. **213**(2), 455 (2009). https://doi.org/10.1016/j.amc.2009.03.037
9. Ramos, T.R.P., Gomes, M.I., Barbosa-Póvoa, A.P.: Omega **48**, 60 (2014). https://doi.org/10.1016/j.omega.2013.11.006
10. Marques, C.M., Moniz, S., de Sousa, J.P., Barbosa-Póvoa, A.P.: Comput. Chem. Eng. **106**, 796 (2017). https://doi.org/10.1016/j.compchemeng.2017.04.008. ESCAPE-26

The Shortest Path in Signed Graphs

Inês Serôdio Costa, Rosa Figueiredo, and Cristina Requejo

Abstract This paper addresses the shortest path problem in a signed graph. Signed graphs are suitable for representing positive/trust and negative/mistrust relationships among the various entities (vertices) of a network. The shortest path in a signed graph can be used to understand how successive relations, even if distant, affect the dynamics of the network. More precisely, the idea is to understand how the relation between any two entities is affected when connected through a signed shortest path. We describe ILP models to obtain positive and negative shortest paths in a signed graph between all pairs of vertices. We evaluate the ILP models on social network benchmark instances and present computational results. Our results highlight potential research opportunities and challenges for the social network optimization community.

Keywords Integer linear programming models · Social networks · Signed paths · Flow models

1 Introduction

Signed graphs are suitable objects to model connections having some positive/trust or negative/mistrust relation among the various entities (vertices) of a network. Initially introduced [2, 8] to represent feelings among people belonging to the same social

I. S. Costa
Department of Mathematics, University of Aveiro, 3810-193 Aveiro, Portugal
e-mail: inesserodiocosta@ua.pt

R. Figueiredo
Laboratoire Informatique d'Avignon, LIA, Avignon Université, 84140 Avignon, France
e-mail: rosa.figueiredo@univ-avignon.fr

C. Requejo (✉)
Center for Research and Development in Mathematics and Applications (CIDMA), Department of Mathematics, University of Aveiro, 3810-193 Aveiro, Portugal
e-mail: crequejo@ua.pt

J. P. Almeida et al. (eds.), *Operational Research*, Springer Proceedings in Mathematics & Statistics 411,
https://doi.org/10.1007/978-3-031-20788-4_4

group, signed graphs were later used to model other systems [3, 4, 7], such as biological networks, international relations networks, risk management networks. The common point of the various applications modeled on signed graphs is the presence of a polarized environment and the willingness to consider it explicitly.

In a signed graph, we associate with each connection (edge/arc) a sign, positive to represent trust relationship, or negative to represent mistrust relationship. Signed graphs can be used to map the extended relations between the various entities represented in the graph. Based on the analysis of different elements of the signed graph (vertices, signed edges/arcs, signed paths, signed cycles, partitions, etc.), it is possible to extract valuable information about the relationships between the vertices as well as about the (distant) relations between the various entities represented by these vertices. The extracted information allows to evaluate the state of the existing relations between the vertices and the behaviour and interconnection between the several entities and the network.

Studying the shortest path problem in a signed graph allows us to have a better knowledge on the extended relationships of the represented entities, particularly those that take into account the positive or negative relation associated with their connection. The idea is to understand how the relationship between two entities is affected when connected through a signed shortest path. With the knowledge of the signed shortest path one can understand to what extent a reliable relationship can be in a network. For example, it can be used to understand how information can be disseminated through entities as a positive/negative information, which means that the information arrives positive if it reaches an entity as it has been disclosed, and the information arrives negative if it reaches an entity contrary to the disclosed information.

We are interested in finding the signed shortest path connection between every pair of vertices. Based on the sign of the shortest path, we intend to study how the influence can be disseminated between pairs of vertices of the signed graph. On the one hand, the influence transmitted from a given origin to a given destination through a positive signed path (a path with an even number of negative signals) arrives at the destination as it has been disclosed. On the other hand, the influence transmitted through a negative signed path (a path with an odd number of negative signals) arrives changed at the destination. With these premises in mind, we study the problem of the shortest path in signed graphs. The objective is to determine how each element represented in the graph is able to influence other elements of the graph, positively or negatively, and at what "speed" that influence occurs. The type of influence is defined by the sign of the signed path, the "speed" is defined by the length of the signed path.

The problem of the shortest path in a signed graph was first posed by Hansen [6] who developed an algorithm, based on the Djikstra shortest path algorithm, to simultaneously find the shortest positive and negative, not necessarily elementary, paths between one vertex and all the other vertices of a signed graph. The algorithm was named the Double Label Algorithm (DLA). A signed shortest path obtained by this DLA algorithm may contain a single circuit with negative sign. If it is required that the obtained paths have no circuits, that is the paths are elementary paths, the algorithm is unable to obtain such paths. Hansen [6] proves that the problem of the shortest elementary signed path is NP-hard. The DLA algorithm was explored by Klamt and Kamp [9] and used in some applications from Biology. The enhanced algorithm from Klamt and Kamp [9] was unable to solve some of the considered instances.

Our contribution in this paper to the shortest path problem in a signed graph is the proposal of integer linear programming (ILP) formulations. We compare different ILP formulations to this problem and report computational experience for the resolution of instances up to 29 vertices.

2 The Shortest Path Problem in Signed Graphs

A signed graph is a graph $G = (V, A)$, $|V| = n$, with a sign $\sigma_{ij} \in \{-1, +1\}$ associated to each arc $(i, j) \in A$, with $A \subseteq \{(i, j) : i, j \in V, i \neq j\}$. A signed graph can be weighted according to the problem we are dealing with. Let $A^+ = \{(i, j) \in A \mid \sigma_{ij} = +1\}$ and $A^- = \{(i, j) \in A \mid \sigma_{ij} = -1\}$ such that $A = A^+ \cup A^-$ and $A^+ \cap A^- = \emptyset$. We consider that associated to each arc $(i, j) \in A$ there is a cost $c_{ij} \in \mathbb{R}_0^+$.

An elementary path $P = [i_1, i_2, \ldots, i_k]$ is a sequence of different vertices of V such that $(i_\ell, i_{\ell+1}) \in A$, $\forall \ell = 1, \ldots, k-1$, and $i_\ell \neq i_{\bar{\ell}}$ for $\ell \neq \bar{\ell}$. Vertex $s = i_1$ is the source vertex of the path, vertex $i_k = t$ is the terminal vertex of the path. We consider elementary shortest paths. The cost $C(P)$ of a path P is the sum of the costs of the arcs in the path. In a signed graph G, we define a *positive path* as a path with an even number of negative arcs and define a *negative path* as a path with an odd number of negative arcs. A *shortest positive path* between a pair of vertices is a positive path with the lowest cost. A *shortest negative path* between a pair of vertices is a negative path with the lowest cost. Notice that if vertex i is in the shortest path (positive/negative) from vertex s to vertex t the subpath between vertices s and i is a shortest path from s to i and may be a positive or a negative shortest path. A *negative cycle* is a cycle with an odd number of negative arcs. A (non-elementary) shortest signed (positive or negative) path may have a negative cycle. Negative cycles can change the sign of a path and can be used to find a shortest negative path between two vertices that are only connected by positive paths or to find a shortest positive path between two vertices that are only connected by negative paths. We want to obtain elementary shortest signed paths, therefore all the shortest signed paths must be obtained without negative cycles.

We aim to find a shortest positive path and a shortest negative path in a signed graph linking any pair of vertices and we want these paths to be elementary. The signed elementary shortest path problem is NP-hard, as the NP-complete problem that decides on the existence of an even path (a path with an even number of arcs) connecting two specific vertices in a graph can be reduced to it (see [6]).

3 Models for the Shortest Path Problem in Signed Graphs

To model the shortest path problem in a signed graph as an integer linear program (ILP) we use two sets of constraints: one modeling the path P from a source vertex to a terminal vertex and the other modeling the sign of the path P. The general model is as follow

$$\begin{aligned} \min \quad & C(P) \\ s.t. \quad & P \in X, \end{aligned} \tag{1}$$

$$P \in S. \tag{2}$$

The objective function is to minimize the overall cost $C(P)$ of the obtained signed path. The set of constraints $P \in X$ models paths from a source vertex to a terminal vertex. The set of constraints $P \in S$ models the sign of the path. To define the set $P \in X$, a set of topological variables associated to each arc $(i, j) \in A$ is used to define the path, indicating whether the arc is selected to be in the solution. We will describe three alternative models: the first models the shortest path from a specified source vertex $s \in V$ to a specified terminal vertex $t \in V \backslash \{s\}$; the second models the shortest paths from a specified source vertex $s \in V$ to every terminal vertex $t \in V \backslash \{s\}$; and the third models the shortest paths from every source vertex $s \in V$ to every terminal vertex $t \in V \backslash \{s\}$. For each one of the three alternatives we describe the corresponding ILP model in the following three subsections: Sects. 3.1, 3.2 and 3.3. The models we describe for the shortest path problem follow flow formulations and multicommodity flow formulations described in Ahuja, Magnanti and Orlin [1] and in Magnanti and Wolsey [10]. To define the set $P \in S$, we propose two alternative models: one using integer variables indicating the sign of the obtained path, another using binary variables indicating the sign of the path arriving at each vertex. These two alternatives are discussed in Sect. 3.4. The last subsection, Sect. 3.5, introduces well known cut inequalities used to eliminate (negative) cycles from solutions and allow to obtain elementary shortest signed paths.

3.1 Model for the Shortest Path Problem from Vertex s to Vertex t

Consider the shortest path problem from a specific vertex s to a specific vertex t, with $s, t \in V, s \neq t$. Consider a set of topological binary variables x_{ij} associated to each arc $(i, j) \in A$, indicating whether arc (i, j) is selected to be in the solution ($x_{ij} = 1$), or arc (i, j) is not selected to be in the solution ($x_{ij} = 0$). Define $C(P) = \sum_{(i,j) \in A} c_{ij} x_{ij}$. An integer linear programming (ILP) model for the set of constraints $P \in X$ using the usual flow conservation constraints follows:

$$\sum_{i \in V} x_{si} = 1, \tag{3}$$

$$\sum_{i \in V} x_{ij} - \sum_{i \in V} x_{ji} = 0, \qquad j \in V \setminus \{s, t\}, \tag{4}$$

$$\sum_{i \in V} x_{it} = 1, \tag{5}$$

$$x_{ij} \in \{0, 1\}, \qquad (i, j) \in A. \tag{6}$$

3.2 *Model for the Shortest Path Problem from Vertex s to Every Vertex $t \in V \setminus \{s\}$*

Consider the shortest path problem from a specific vertex $s \in V$, to every other vertices $t \in V \setminus \{s\}$. Using the set of topological binary variables y^t_{ij}, defined for each arc $(i, j) \in A$ and each terminal vertex $t \in V \setminus \{s\}$, indicating whether arc (i, j) is selected to be in the solution ($y^t_{ij} = 1$), or arc (i, j) is not selected to be in the solution ($y^t_{ij} = 0$). Define the paths cost as $C(P) = \sum_{t \in V} \sum_{(i,j) \in A} c_{ij} y^t_{ij}$. An integer linear programming (ILP) model for the problem using the usual multicommodity flow conservation constraints follows:

$$\sum_{i \in V} y^t_{si} = 1, \qquad t \in V \setminus \{s\}, \tag{7}$$

$$\sum_{i \in V} y^t_{ij} - \sum_{i \in V} y^t_{ji} = 0, \qquad j \in V \setminus \{s, t\}, t \in V \setminus \{s\}, \tag{8}$$

$$\sum_{i \in V} y^t_{it} = 1, \qquad t \in V \setminus \{s\}, \tag{9}$$

$$y^t_{ij} \in \{0, 1\}, \qquad (i, j) \in A, t \in V \setminus \{s\}. \tag{10}$$

3.3 *Model for the Shortest Path Problem from Every Vertex $s \in V$ to Every Vertex $t \in V \setminus \{s\}$*

For every pair of source and terminal vertices (s, t), $s, t \in V$, $s \neq t$, consider the shortest path problem from a source vertex s to a terminal vertex t. Using the set of topological binary variables z^{st}_{ij}, defined for each arc $(i, j) \in A$ and each pair of source and terminal vertices $s, t \in V$, $s \neq t$, indicating whether arc (i, j) is selected to be in the solution ($z^{st}_{ij} = 1$), or arc (i, j) is not selected to be in the solution ($z^{st}_{ij} = 0$). Define the paths cost as $C(P) = \sum_{s,t \in V, s \neq t} \sum_{(i,j) \in A} c_{ij} z^{st}_{ij}$. An integer linear programming (ILP) model for the problem using the usual multicommodity flow conservation constraints follows:

$$\sum_{i \in V} z_{si}^{st} = 1, \qquad s, t \in V, s \neq t, \tag{11}$$

$$\sum_{i \in V} z_{ij}^{st} - \sum_{i \in V} z_{ji}^{st} = 0, \qquad j \in V \backslash \{s, t\}, s, t \in V, s \neq t, \tag{12}$$

$$\sum_{i \in V} z_{it}^{st} = 1, \qquad s, t \in V, s \neq t, \tag{13}$$

$$z_{ij}^{st} \in \{0, 1\}, \qquad (i, j) \in A, s, t \in V, s \neq t. \tag{14}$$

3.4 *Models to Define the Sign of the Path*

Next, we describe two alternative sets of constraints $P \in S$ for establishing the sign of the path. One associates a variable to each signed path, defined to specify the sign of the obtained path. The other associates a variable to each vertex indicating the sign of the path arriving at that vertex. For each, we consider first the shortest path problem from a vertex s to a vertex t and then describe small changes to accommodate for the other shortest path models.

In the first set of constraints, we use a unique variable $q \in \mathbb{N}$ associated to the path that counts the number of arcs in the path with a negative sign. When the number of arcs with a negative sign is even the path is positive, when it is odd the path is negative. Therefore, when set S is defined by the constraint

$$\sum_{(i,j) \in A^-} x_{ij} = 2q \tag{15}$$

a positive shortest path is obtained, having $2q$ arcs of negative sign. When set S is defined by the constraint

$$\sum_{(i,j) \in A^-} x_{ij} = 2q - 1 \tag{16}$$

a negative shortest path is obtained, having $2q - 1$ arcs of negative sign.

We name model using constraints (3)–(6) together with one of the constraints (15) or (16) as model ST-1.

To obtain a model using constraints (7)–(10), a variable q^t is associated to each path from vertex s to every vertex $t \in V \backslash \{s\}$. Then the constraint (15) or (16) should be accordingly written using variables y_{ij}^t and replacing variable q for variables q^t, hence obtaining a constraint for each path to vertex $t \in V \backslash \{s\}$. We obtain a set of $|V| - 1$ constraints for each option: obtain positive paths using constraints (15) or obtain negative paths using constraints (16). We name model using constraints (7)–(10) together with one set of the rewritten and extended constraints (15) or (16) as model SA-1.

To obtain a model using constraints (11)–(14), a variable q^{st} is associated to each path from every vertex $s \in V$ to every vertex $t \in V \backslash \{s\}$. Then the constraints (15) or (16) should be accordingly written using variables z_{ij}^{st}. We obtain a set of $|V| \times (|V| - 1)$ constraints for each option: obtain positive paths or obtain negative paths. We name model using constraints (11)–(14) together with one set of the rewritten and extended constraints (15) or (16) as model AA-1,

In the second set of constraints, we use binary variables $p_i \in \{0, 1\}$, for all $i \in V$, indicating whether the path arriving at vertex i is a positive path ($p_i = 1$) or a negative path ($p_i = 0$). The following set of constraints define variables $p_i \in \{0, 1\},\ i \in V$.

$$p_j \geq x_{ij} - p_i, \qquad (i, j) \in A^- \tag{17}$$

$$p_j \leq 2 - x_{ij} - p_i, \qquad (i, j) \in A^- \tag{18}$$

$$p_j \geq x_{ij} + p_i - 1, \qquad (i, j) \in A^+ \tag{19}$$

$$p_j \leq 1 - x_{ij} + p_i, \qquad (i, j) \in A^+ \tag{20}$$

Further, initialize the variable p_s at the starting vertex s of the path with $p_s = 1$. Additionally, if path to vertex t must be positive set $p_t = 1$. If path to vertex t must be negative set $p_t = 0$. Notice that, in the case we do not set p_t, we obtain a shortest path which is a positive path if we get $p_t = 1$, and is a negative path if we get $p_t = 0$.

We name model using constraints (3)–(6) together with constraints (17)–(20) as model ST-2.

To obtain a model using constraints (7)–(10) a variable p_i^t is associated to each vertex $i \in V$, for each path from vertex s to every vertex $t \in V \backslash \{s\}$. Then the constraints (17)–(20) are accordingly written using variables y_{ij}^t and by replacing variables p_i for variables p_i^t. We obtain a set of constraints for each path to vertex $t \in V \backslash \{s\}$. We name model using constraints (7)–(10) together with the rewritten and extended constraints (17)–(20) as model SA-2.

To obtain a model using constraints (11)–(14) a variable p_i^{st} is associated to each vertex $i \in V$, for each path from every vertex $s \in V$ to every vertex $t \in V \backslash \{s\}$. Then the constraints (17)–(20) are accordingly written using variables z_{ij}^{st} and by replacing variables p_i for variables p_i^{st}. We obtain a set of constraints for each path from every vertex $s \in V$ to every vertex $t \in V \backslash \{s\}$. We name model using constraints (11)–(14) together with the rewritten and extended constraints (17)–(20) as model AA-2.

3.5 Eliminate Negative Cycles

When using the models ST-1, SA-1 and AA-1 the obtained paths may have a negative cycle, thus the models obtain positive or negative shortest paths that are non elementary paths. To guaranty that a negative cycle is not used in the obtained paths, cycle or subtour elimination constraints must be included in the models (see [11]). For any subset $Q \subseteq V$ of the set of vertices, $|Q| \geq 2$, the well known subtour elimination constraints

$$\sum_{i,j \in Q} x_{ij} \leq |Q| - 1 \tag{21}$$

eliminate any such cycles. These constraints are in exponential number, thus in this study we consider a small set of these inequalities for sets Q of size 2 and 3. We adapt these constraints for each topological set of variables y and z, and we name the new models ST-1*, SA-1* and AA-1*, respectively, to the models ST-1, SA-1 and AA-1 with the inequalities (21) added for sets Q of size 2 and 3.

4 Computational Tests

The ILP models were implemented in Mosel and computational tests were run using XpressMP 8.5 with the default options on a computer with an Intel(R) Core(TM) i7-4750HQ processor, 2.00GHz CPU, and 8GB of RAM. We consider 21 test instances from [5]. Additionally, just for reference, we also use the small example with 6 vertices from [6], named Hansen. The 21 test instances from [5] are non-complete graphs of medium size having a number of vertices ranging from 17 to 21. All instances, but one, are directed and the number of negative arcs is similar to the number of positive arcs. The instance named *McKinn* (does not follow these rules) it is not directed and has more positive edges (246) than negative edges (18). Having such a few number of negative edges turns the instance difficult when obtaining negative shortest paths (between all pairs of vertices). Notice that these instances are not complete and that the total number of arcs of each instance (which is $|A^+| + |A^-|$) is lower than the total number of arcs of a complete graph (which is M).

We obtained computational results using the six combinations of the three shortest path models with the two signal path models. As we want to obtain shortest paths between every pair of vertices, we proceed as follows with each model. We run the models ST-1, ST-1* and ST-2 that model the shortest path from a specified vertex s to a specified vertex t, for every $s \in V$ and every $t \in V\backslash\{s\}$, therefore we run the models $|V| \times (|V| - 1)$ times. We run the models SA-1, SA-1* and SA-2 that model the shortest path from a specified vertex s to every vertex $t \in V\backslash\{s\}$, for every $s \in V$ therefore we run the models $|V|$ times. We run the models AA-1, AA-1* and AA-2 that model the shortest path from every vertex $s \in V$ to every vertex $t \in V\backslash\{s\}$, therefore we run the models one time.

With each one of the six models we obtain the following. (i) We obtain a shortest positive path between every pair of vertices, if a positive path exists. These results are reported in Table 1. (ii) We obtain a shortest negative path between every pair of vertices, if a negative path exists. These results are reported in Table 2. Additionally (iii) we obtain a shortest path, if a path exists, this shortest path may be a positive signed path or a negative signed path, these results are reported in Table 3. The problem of obtaining in (iii) a shortest (signed) path is easy (polynomial), but obtaining in (i) a elementary shortest positive path or in (ii) a elementary shortest negative path are both NP-hard problems.

Using models ST-2, SA-2 and AA-2 we obtain solutions that correspond to (elementary) positive and negative shortest paths. Using models ST-1, SA-1, AA-1,

ST-1*, SA-1* and AA-1*, the positive and negative shortest paths that we obtain can have negative cycles.

Table 1 shows sample results when obtaining all the possible shortest positive paths and Table 2 shows sample results when obtaining all the possible shortest negative paths. For better layout each table is divided in three parts. The top part shows results obtained by the models ST-1, SA-1 and AA-1 (thus without the cycle constraints (21)), and therefore possibly obtaining non elementary paths. The middle part of the table shows results obtained by the models ST-1*, SA-1* and AA-1* with the constraints (21) included only for sets Q of size 2 and 3, thus it is still possible that some non-elementary paths can be obtained. The bottom part of the tables shows results obtained by the models ST-2, SA-2 and AA-2.

In each part, the first seven columns show the information about the problem instance: name of the instance, number of vertices $|V|$, number of pairs of vertices $M = (|V| - 1) \times |V|$, number of arcs with positive sign $|A^+|$, number of arcs with negative sign $|A^-|$, the value C of the optimum cost of a shortest positive/negative path obtained between all pairs of vertices, the total number of positive/negative shortest paths obtained in the optimal solution. Next, for each ILP model identified in the first line, the respective columns show: the value C for the overall cost of shortest positive/negative paths obtained, the total number of positive/negative paths obtained, and the computational times (in seconds) used by each ILP model to obtain the two previous values. When solving the problem a time limit of 3600 s (1 h) was imposed. For some instances and models we were not able to obtain solutions within the time limit, these results are marked with an asterisk * in place of the computational time used (in such cases we report the cost and the number of paths obtained within that time limit). Notice that, for each instance and each model it is displayed in Table 1 the number of positive shortest paths and in Table 2 the number of negative shortest paths that is possible to obtain between all pairs of vertices, these values are reported in columns named $|P^+|$ and $|P^-|$ (respectively).

The cost values obtained using models ST-1, SA-1 and AA-1 (displayed in the top part of Tables 1 and 2) and models ST-1*, SA-1* and AA-1* (displayed in the middle part of Tables 1 and 2) represent lower bounds on the optimal solution value when the number of shortest signed paths obtained is equal to the number of shortest signed paths of the optimal solution. For example, consider instance *Monk2*. Using models ST-1, SA-1 and AA-1, the obtained cost of the shortest paths is 1042 for 306 positive paths (see Table 1, first part). The number of shortest positive paths in the optimal solution is also 306 thus the cost value of 1042 represents a lower bound on the optimal cost value. The cost values do not represent lower bounds when the number of shortest signed paths obtained differs from this number in their optimal solution. This means that the obtained solution may have some non-elementary paths (if having some negative cycles). These cases are marked with an asterisk * next to the number of paths obtained that differ from the number of the optimal solution. When these values (cost and number of paths) are greater than the corresponding values in the optimal solution we may conclude that it is possible that a non-elementary shortest signed path was obtained for some pairs of vertices that do not have a shortest signed path. This is the case of instance *bddate*. Using models ST-1, SA-1 and AA-

Table 1 Computational results for models when obtaining shortest positive paths

Name	$\|V\|$	$\|M\|$	$\|A^+\|$	$\|A^-\|$	C	$\|P^+\|$	ST-1			SA-1			AA-1		
							C	$\|P^+\|$	Time	C	$\|P^+\|$	Time	C	$\|P^+\|$	Time
Hansen	6	30	4	6	147	16	147	16	0.0	147	16	0.0	147	16	0.0
b_sum	17	272	78	83	864	272	851	272	1.3	851	272	1.6	851	272	3.8
bddate	17	272	51	51	562	271	565	*272	1.3	565	*272	1.3	565	*272	4.9
bfriend	17	272	51	51	571	272	570	272	1.3	570	272	1.8	570	272	7.1
broomm	17	272	51	51	573	272	573	272	1.4	573	272	1.4	573	272	2.2
bweeke	17	272	51	51	573	272	571	272	1.4	571	272	1.9	571	272	5.4
NewCmb	17	272	68	51	516	272	516	272	1.1	516	272	1.0	516	272	2.2
Monk2	18	306	55	49	1052	306	1042	306	1.6	1042	306	2.1	1042	306	3.8
Monk3	18	306	57	48	1081	306	1046	306	1.5	1046	306	1.4	1046	306	2.8
Monk4	18	306	56	47	1153	306	1130	306	1.9	1130	306	3.3	1130	306	19.5
Monk4S	18	306	78	76	1299	306	1276	306	1.5	1276	306	1.6	1276	306	2.2
c_sum	20	380	93	108	1150	380	1133	380	2.6	1133	380	4.3	1133	380	53.0
cddate	20	380	60	60	914	380	910	380	2.9	910	380	4.5	910	380	11.0
cfriend	20	380	60	60	796	344	790	344	3.2	790	344	5.8	790	344	37.7
croomm	20	380	60	60	886	380	882	380	2.7	882	380	4.2	882	380	13.0
cweeke	20	380	60	60	889	380	888	380	3.8	888	380	5.1	888	380	11.5
a_sum	21	420	92	119	1173	420	1155	420	2.8	1155	420	5.3	1155	420	29.4
addate	21	420	63	63	967	420	962	420	3.2	962	420	5.2	962	420	13.2
afriend	21	420	63	63	1009	419	1009	*420	3.1	1009	*420	4.6	1009	*420	11.0
aroomm	21	420	63	63	1021	419	1012	*420	3.3	1012	*420	4.3	1012	*420	43.7
aweeke	21	420	63	63	980	420	972	420	3.1	972	420	4.2	972	420	13.1
McKinn	29	812	246	18	1474	812	1474	812	8.2	1474	812	5.6	1474	812	14.0

(continued)

Table 1 (continued)

Name	$\|V\|$	$\|M\|$	$\|A^+\|$	$\|A^-\|$	C	$\|P^+\|$	ST-1*			SA-1*			AA-1*		
							C	$\|P^+\|$	Time	C	$\|P^+\|$	Time	C	$\|P^+\|$	Time
Hansen	6	30	4	6	147	16	147	16	0.0	147	16	0.0	147	16	0.0
b_sum	17	272	78	83	864	272	864	272	14.6	864	272	12.7	864	272	38.1
bddate	17	272	51	51	562	271	567	*272	13.4	567	*272	8.5	567	*272	18.4
bfriend	17	272	51	51	571	272	571	272	13.5	571	272	9.2	571	272	36.2
broomm	17	272	51	51	573	272	573	272	13.3	573	272	8.3	573	272	19.6
bweeke	17	272	51	51	573	272	573	272	13.5	573	272	9.2	573	272	33.8
NewCmb	17	272	68	51	516	272	516	272	13.0	516	272	7.4	516	272	13.1
Monk2	18	306	55	49	1052	306	1049	306	18.4	1049	306	12.8	1049	306	34.4
Monk3	18	306	57	48	1081	306	1075	306	18.2	1075	306	12.0	1075	306	78.2
Monk4	18	306	56	47	1153	306	1149	306	19.5	1149	306	16.3	1150	306	*
Monk4S	18	306	78	76	1299	306	1295	306	18.9	1295	306	13.7	1295	306	81.5
c_sum	20	380	93	108	1150	380	1147	380	40.9	1147	380	30.1	1147	380	*
cddate	20	380	60	60	914	380	914	380	38.8	914	380	25.7	914	380	446.6
cfriend	20	380	60	60	796	344	796	344	38.7	796	344	25.3	796	344	*
croomm	20	380	60	60	886	380	886	380	37.5	886	380	22.5	887	380	*
cweeke	20	380	60	60	889	380	889	380	39.6	889	380	23.9	889	380	87.0
a_sum	21	420	92	119	1173	420	1173	420	55.5	1173	420	41.6	1176	420	*
addate	21	420	63	63	967	420	967	420	53.5	967	420	29.6	967	420	3220
afriend	21	420	63	63	1009	419	1014	*420	51.9	1014	*420	27.4	1014	*420	152.2
aroomm	21	420	63	63	1021	419	1026	*420	53.3	1026	*420	56.8	1027	*420	*
aweeke	21	420	63	63	980	420	980	420	52.3	980	420	32.3	980	420	*
McKinn	29	812	246	18	1474	812	1474	812	434.8	1474	812	110	1474	812	357.0

(continued)

Table 1 (continued)

Name	$\|V\|$	$\|M\|$	$\|A^+\|$	$\|A^-\|$	C	$\|P^+\|$	ST-2			SA-2			AA-2		
							C	$\|P^+\|$	Time	C	$\|P^+\|$	Time	C	$\|P^+\|$	Time
Hansen	6	30	4	6	147	16	147	16	0.0	147	16	0.0	147	16	0.0
b_sum	17	272	78	83	864	272	864	272	2.0	864	272	2.6	864	272	3.6
bddate	17	272	51	51	562	271	562	271	1.8	562	271	2.2	562	271	3.1
bfriend	17	272	51	51	571	272	571	272	1.8	571	272	2.1	571	272	2.8
broomm	17	272	51	51	573	272	573	272	1.8	573	272	2.1	573	272	3.9
bweeke	17	272	51	51	573	272	573	272	1.8	573	272	2.1	573	272	3.0
NewCmb	17	272	68	51	516	272	516	272	1.8	516	272	2.1	516	272	2.7
Monk2	18	306	55	49	1052	306	1052	306	2.3	1052	306	2.9	1052	306	6.1
Monk3	18	306	57	48	1081	306	1081	306	2.3	1081	306	3.2	1081	306	6.8
Monk4	18	306	56	47	1153	306	1153	306	2.6	1153	306	4.4	1153	306	15.5
Monk4S	18	306	78	76	1299	306	1299	306	2.4	1299	306	3.1	1299	306	4.6
c_sum	20	380	93	108	1150	380	1150	380	3.7	1150	380	5.2	1150	380	15.8
cddate	20	380	60	60	914	380	914	380	3.7	914	380	6.1	914	380	68.5
cfriend	20	380	60	60	796	344	796	344	3.8	796	344	7.1	796	344	2299
croomm	20	380	60	60	886	380	886	380	3.3	886	380	4.5	886	380	10.8
cweeke	20	380	60	60	889	380	889	380	3.7	889	380	5.2	889	380	23.2
a_sum	21	420	92	119	1173	420	1173	420	4.5	1173	420	5.9	1173	420	9.6
addate	21	420	63	63	967	420	967	420	4.2	967	420	5.6	967	420	46.7
afriend	21	420	63	63	1009	419	1009	419	4.6	1009	419	7.0	1009	419	48.8
aroomm	21	420	63	63	1021	419	1021	419	4.6	1021	419	8.8	1021	419	81.4
aweeke	21	420	63	63	980	420	980	420	4.2	980	420	5.3	980	420	13.7
McKinn	29	812	246	18	1474	812	1474	812	14.4	1474	812	14.0	1474	812	31.5

Table 2 Computational results for models when obtaining shortest negative paths

Name	$\|V\|$	$\|M\|$	$\|A^+\|$	$\|A^-\|$	C	$\|P^-\|$	ST-1			SA-1			AA-1		
							C	$\|P^-\|$	Time	C	$\|P^-\|$	Time	C	$\|P^-\|$	Time
Hansen	6	30	4	6	137	17	137	17	0.1	137	17	0.0	137	17	0.0
b_sum	17	272	78	83	851	272	838	272	1.3	838	272	1.2	838	272	1.8
bddate	17	272	51	51	569	272	569	272	1.2	569	272	1.4	569	272	3.0
bfriend	17	272	51	51	579	272	578	272	1.5	578	272	1.6	578	272	3.5
broomm	17	272	51	51	575	272	573	272	1.3	573	272	1.5	573	272	4.4
bweeke	17	272	51	51	570	272	569	272	1.2	569	272	1.5	569	272	3.1
NewCmb	17	272	68	51	547	272	547	272	1.2	547	272	1.3	547	272	1.7
Monk2	18	306	55	49	1087	305	1081	*306	1.7	1081	*306	2.3	1081	*306	5.7
Monk3	18	306	57	48	1093	305	1056	*306	1.4	1056	*306	1.3	1056	*306	2.9
Monk4	18	306	56	47	1162	306	1135	306	1.9	1135	306	4.6	1135	306	37.5
Monk4S	18	306	78	76	1379	306	1322	306	1.5	1322	306	1.9	1322	306	4.6
c_sum	20	380	93	108	1108	376	1103	*377	2.5	1103	*377	2.9	1103	*380	7.1
cddate	20	380	60	60	899	379	893	*380	2.6	893	*380	4.7	893	*380	9.3
cfriend	20	380	60	60	769	342	769	*344	3.0	769	*344	4.4	769	*344	13.4
croomm	20	380	60	60	854	379	852	*380	2.6	852	*380	3.8	852	*380	8.2
cweeke	20	380	60	60	902	379	897	*380	3.7	897	*380	5.6	897	*380	13.2
a_sum	21	420	92	119	1130	420	1123	420	2.7	1123	420	3.4	1123	420	9.1
addate	21	420	63	63	951	420	945	420	2.8	945	420	4.0	945	420	10.7
afriend	21	420	63	63	975	420	967	420	2.9	967	420	3.8	967	420	7.9
aroomm	21	420	63	63	999	419	988	*420	2.8	988	*420	2.9	988	*420	9.5
aweeke	21	420	63	63	972	420	966	420	2.7	966	420	4.3	966	420	7.1
McKinn	29	812	246	18	2434	812	2430	812	11.1	2430	812	17.9			*

(continued)

Table 2 (continued)

Name	$\|V\|$	$\|M\|$	$\|A^+\|$	$\|A^-\|$	C	$\|P^-\|$	ST-1*			SA-1*			AA-1*		
							C	$\|P^-\|$	Time	C	$\|P^-\|$	Time	C	$\|P^-\|$	Time
Hansen	6	30	4	6	137	17	137	17	0.0	137	17	0.0	137	17	0.0
b_sum	17	272	78	83	851	272	851	272	14.5	851	272	10.1	851	272	20.4
bddate	17	272	51	51	569	272	569	272	13.2	569	272	7.9	569	272	11.2
bfriend	17	272	51	51	579	272	579	272	14.2	579	272	9.2	579	272	13.7
broomm	17	272	51	51	575	272	575	272	13.4	575	272	8.1	575	272	14.0
bweeke	17	272	51	51	570	272	570	272	13.4	570	272	8.8	570	272	14.2
NewCmb	17	272	68	51	547	272	547	272	13.5	547	272	7.5	547	272	11.5
Monk2	18	306	55	49	1087	305	1092	*306	19.5	1092	*306	14.2	1092*	306	59.4
Monk3	18	306	57	48	1093	305	1090	*306	18.5	1090	*306	11.7	1090*	306	35.7
Monk4	18	306	56	47	1162	306	1159	306	19.6	1159	306	19.6	1161	306	*
Monk4S	18	306	78	76	1379	306	1366	306	20.2	1366	306	28.2	1369	306	*
c_sum	20	380	93	108	1108	376	1114	*377	40.2	1114	*377	24.9	1114	*380	155.4
cddate	20	380	60	60	899	379	904	*380	39.9	904	*380	28.7	904	*380	*
cfriend	20	380	60	60	769	342	779	*344	39.3	779	*344	23.5	779	*344	*
croomm	20	380	60	60	854	379	859	*380	38.7	859	*380	22.1	859	*380	1062
cweeke	20	380	60	60	902	379	907	*380	39.8	907	*380	27.7	909	*380	*
a_sum	21	420	92	119	1130	420	1130	420	53.4	1130	420	27.8	1130	420	156.1
addate	21	420	63	63	951	420	951	420	50.3	951	420	25.6	951	420	812.1
afriend	21	420	63	63	975	420	975	420	51.0	975	420	26.1	975	420	583.7
aroomm	21	420	63	63	999	419	1004	*420	52.7	1004	*420	37.5	1004	*420	*
aweeke	21	420	63	63	972	420	972	420	50.4	972	420	24.6	972	420	621.0
McKinn	29	812	246	18	2434	812			3600*			3600*			*

(continued)

Table 2 (continued)

Name	$\|V\|$	$\|M\|$	$\|A^+\|$	$\|A^-\|$	C	$\|P^-\|$	ST-2			SA-2			AA-2		
							C	$\|P^-\|$	Time	C	$\|P^-\|$	Time	C	$\|P^-\|$	Time
Hansen	6	30	4	6	137	17	137	17	0.0	137	17	0.0	137	17	0.0
b_sum	17	272	78	83	851	272	851	272	1.9	851	272	2.5	851	272	3.3
bddate	17	272	51	51	569	272	569	272	1.7	569	272	2.1	569	272	2.9
bfriend	17	272	51	51	579	272	579	272	1.7	579	272	2.4	579	272	5.6
broomm	17	272	51	51	575	272	575	272	1.8	575	272	2.0	575	272	4.3
bweeke	17	272	51	51	570	272	570	272	1.8	570	272	2.1	570	272	3.1
NewCmb	17	272	68	51	547	272	547	272	1.8	547	272	2.1	547	272	2.7
Monk2	18	306	55	49	1087	305	1087	305	2.4	1087	305	3.6	1087	305	10.0
Monk3	18	306	57	48	1093	305	1093	305	2.2	1093	305	2.6	1093	305	4.1
Monk4	18	306	56	47	1162	306	1162	306	2.7	1162	306	4.4	1162	306	16.3
Monk4S	18	306	78	76	1379	306	1379	306	2.6	1379	306	3.9	1379	306	8.2
c_sum	20	380	93	108	1108	376	1108	376	3.8	1108	376	5.4	1108	376	14.0
cddate	20	380	60	60	899	379	899	379	3.6	899	379	5.0	899	379	26.0
cfriend	20	380	60	60	769	342	769	342	3.6	769	342	5.8	769	342	11.7
croomm	20	380	60	60	854	379	854	379	3.2	854	379	3.8	854	379	7.2
cweeke	20	380	60	60	902	379	902	379	3.8	902	379	5.4	902	379	23.9
a_sum	21	420	92	119	1130	420	1130	420	4.3	1130	420	5.6	1130	420	9.5
addate	21	420	63	63	951	420	951	420	3.9	951	420	4.9	951	420	11.3
afriend	21	420	63	63	975	420	975	420	4.1	975	420	6.1	975	420	31.3
aroomm	21	420	63	63	999	419	999	419	4.1	999	419	4.9	999	419	12.9
aweeke	21	420	63	63	972	420	972	420	3.9	972	420	4.5	972	420	8.2
McKinn	29	812	246	18	2434	812	2434	812	40.8			*		*	

1, the obtained cost of the shortest paths is 565 for 272 positive paths (see Table 1, first part), however the number of shortest positive paths in the optimal solution is 271 therefore the cost value is not a lower bound on the optimal cost value (which is 562). For example, with instance *aroomm*, using models ST-1, SA-1 and AA-1, the obtained cost of the shortest paths is 1012 for 420 positive paths (see Table 1, first part), however the number of shortest positive paths in the optimal solution is 419. Still instance *aroomm*, using models ST-1*, and SA-1*, the obtained cost of the shortest paths is 1026 for 420 positive paths (see Table 1, middle part), which is greater than the number of shortest positive paths in the optimal solution that is 419, therefore the cost value is not a lower bound on the optimal cost value, which is 1021. In both non-optimal solutions, paths with negative cycles were found.

Models ST-1*, SA-1* and AA-1* having the cut constraints (21) for sets of size 2 and 3 became very hard to solve. However, using models ST-2, SA-2 and AA-2 we were able to obtain shortest signed paths using a few seconds for most of the instances. The instance named $McKinn$ proved to be very hard to solve, however we were able to obtain shortest positive paths between every pair of vertices using models ST-2, SA-2 and AA-2 in less than 35 s and we were able to obtain shortest negative paths between every pair of vertices using models ST-2 in less than 50 s.

Just for reference, Table 3 shows sample results for obtaining the shortest paths between all pairs of vertices. This table is divided into two parts. In the first part we show results obtained using models for the sign of the path using variables q and constraint

$$\sum_{(i,j)\in A^-} x_{ij} \leq 2q \tag{22}$$

for describing set S. In the second part we show results obtained using models for the sign of the path using variables p_i and constraints (17)–(20) for describing set S (and not setting a value for variable p_t). In these two cases the shortest path is obtained as we do not set the path to be positive or negative and at the end we identify if the obtained path is a positive or a negative shortest path. For each part, from left to right we show the information about the problem instance (similarly to the previous tables), followed by details of the solution obtained with each ILP model identified in the first line (number of positive paths obtained, number of negative paths obtained, computational time in seconds used to obtain the solution). The number of positive and negative paths that are a shortest path for each instance is displayed, for each model, in columns named $|P^+|$ and $|P^-|$, respectively. The obtained cost value is the cost of all the shortest paths between all pairs of vertices. Notice that when obtaining the shortest path in case of existence of alternative optimal shortest path positive and negative the number of positive and negative paths obtained by each model may be different. This number is not the same for all the models as in most cases there are alternative paths. However it is smaller than or equal to the number M of pairs of vertices. For some instances, for instance, for the smaller instance named *Hansen* this number is less than M meaning that for some pairs of vertices there is no path at all connecting the two vertices. The computational results show that the solutions to all instances were obtained quite quickly using only a few seconds of computational

Table 3 Computational results for models when obtaining the shortest (signed) path, either positive or negative

Name	$\|V\|$	$\|M\|$	$\|A^+\|$	$\|A^-\|$	C	ST-1			SA-1			AA-1		
						$\|P^+\|$	$\|P^-\|$	Time	$\|P^+\|$	$\|P^-\|$	Time	$\|P^+\|$	$\|P^-\|$	Time
Hansen	6	30	4	6	128	10	11	0.0	10	11	0.0	10	11	0.0
b_sum	17	272	78	83	710	137	135	0.9	142	130	0.6	144	128	0.7
bddate	17	272	51	51	461	134	138	0.9	138	134	0.6	139	133	0.7
bfriend	17	272	51	51	454	138	134	0.9	136	136	0.5	137	135	0.7
broomm	17	272	51	51	456	125	147	0.9	124	148	0.6	128	144	0.7
bweeke	17	272	51	51	454	138	134	0.9	135	137	0.6	142	130	0.7
NewCmb	17	272	68	51	433	158	114	0.8	166	106	0.6	157	115	0.7
Monk2	18	306	55	49	835	161	145	1.0	155	151	0.7	162	144	0.8
Monk3	18	306	57	48	873	164	142	1.0	166	140	0.7	167	139	0.8
Monk4	18	306	56	47	871	142	164	1.0	145	161	0.7	153	153	0.8
Monk4S	18	306	78	76	1066	172	134	1.1	172	134	0.8	166	140	1.0
c_sum	20	380	93	108	940	179	198	1.6	190	187	1.2	192	188	1.6
cddate	20	380	60	60	706	181	199	1.6	178	202	1.0	181	199	1.4
cfriend	20	380	60	60	599	159	185	1.5	158	186	1.0	161	183	1.4
croomm	20	380	60	60	695	164	216	1.6	165	215	1.0	161	219	1.4
cweeke	20	380	60	60	693	180	200	1.5	181	199	1.1	183	197	1.3
a_sum	21	420	92	119	948	204	216	2.0	209	211	1.4	206	214	1.9
addate	21	420	63	63	762	203	217	1.9	199	221	1.2	196	224	1.7
afriend	21	420	63	63	783	198	222	1.9	203	217	1.2	202	218	1.7
aroomm	21	420	63	63	796	212	208	1.9	212	208	1.2	207	213	1.7
aweeke	21	420	63	63	771	204	216	1.9	201	219	1.3	210	210	1.7
McKinn	29	812	246	18	1430	680	132	7.8	655	157	4.4	668	144	8.9

(continued)

Table 3 (continued)

Name	$\|V\|$	$\|M\|$	$\|A^+\|$	$\|A^-\|$	C	ST-2			SA-2			AA-2		
						$\|P^+\|$	$\|P^-\|$	Time	$\|P^+\|$	$\|P^-\|$	Time	$\|P^+\|$	$\|P^-\|$	Time
Hansen	6	30	4	6	128	10	11	0.0	10	11	0.0	10	11	0.0
b_sum	17	272	78	83	710	136	136	1.9	119	153	2.1	111	161	2.8
bddate	17	272	51	51	461	115	157	1.7	109	163	1.8	107	165	2.2
bfriend	17	272	51	51	454	115	157	1.7	109	163	1.7	111	161	2.1
broomm	17	272	51	51	456	117	155	1.7	106	166	1.7	110	162	2.1
bweeke	17	272	51	51	454	115	157	1.7	111	161	1.7	103	169	2.1
NewCmb	17	272	68	51	433	123	149	1.8	111	161	1.8	114	158	2.4
Monk2	18	306	55	49	835	157	149	2.1	152	154	2.1	156	150	2.9
Monk3	18	306	57	48	873	153	153	2.1	150	156	2.1	150	156	3.2
Monk4	18	306	56	47	871	144	162	2.1	143	163	2.1	148	158	3.0
Monk4S	18	306	78	76	1066	163	143	2.2	160	146	2.4	160	146	3.5
c_sum	20	380	93	108	940	169	208	3.5	159	218	3.8	164	216	5.9
cddate	20	380	60	60	706	166	214	3.1	153	227	3.1	153	227	5.2
cfriend	20	380	60	60	599	147	197	3.0	135	209	3.1	130	214	5.1
croomm	20	380	60	60	695	146	234	3.1	125	255	3.1	130	250	5.2
cweeke	20	380	60	60	693	168	212	3.0	162	218	3.1	177	203	4.7
a_sum	21	420	92	119	948	192	228	4.2	177	243	4.4	171	249	6.8
addate	21	420	63	63	762	173	247	3.6	150	270	3.7	156	264	6.0
afriend	21	420	63	63	783	175	245	3.7	151	269	3.7	153	267	6.0
aroomm	21	420	63	63	796	195	225	3.7	172	248	3.7	171	249	6.6
aweeke	21	420	63	63	771	188	232	3.6	166	254	3.7	163	257	6.0
McKinn	29	812	246	18	1430	665	147	14.1	674	138	13.3	685	127	29.1

time. The fastest model to obtain all the possible shortest paths is model SA-1 and, for each instance, all the shortest paths were obtained in less than 10 s, except instance *McKinn* that used less than 30 s.

5 Conclusion

We considered the shortest signed path problem and presented ILP models to obtain the shortest positive paths and the shortest negative paths between all pairs of vertices. Obtaining elementary signed paths is a NP-hard problem. For a set of medium size instances, we were able to obtain shortest positive paths and shortest negative paths between every pair of vertices using one of the proposed models in less than 50 s.

Acknowledgements This research was supported by the Fundação para a Ciência e a Tecnologia (FCT), through the research program PESSOA 2018 - project FCT/5141/13/4/ 2018/S (I.S. Costa and C. Requejo) and by Campus France through the research program PESSOA 2018 - project N° 40821YH (R. Figueiredo). It was also partially supported by the Fundação para a Ciência e a Tecnologia (FCT), within Research Unit CIDMA - project UIDB/04106/2020 (C. Requejo).

References

1. Ahuja, R.K., Magnanti, T.L., Orlin, J.B.: Network Flows: Theory, Algorithms and Applications. Prentice-Hall, Prentice (1993)
2. Cartwright, D., Harary, F.: Structural balance: a generalization of Heiderâs theory. Psychol. Rev. **63**, 277–293 (1956)
3. DasGupta, B., Enciso, G.A., Sontag, E., Zhang, Y.: Algorithmic and complexity results for decompositions of biological networks into monotone subsystems. In: Àlvarez, C., Serna, M. (eds) Experimental Algorithms, pp. 253–264. Springer, Berlin (2006)
4. Doreian, P., Mrvar, A.: Structural balance and signed international relations. J. Soc. Struct. **16** (2015)
5. Figueiredo, R., Moura, G.: Mixed integer programming formulations for clustering problems related to structural balance. Soc. Netw. **35**(4), 639–651 (2013)
6. Hansen, P.: Shortest paths in signed graphs. In: Burkard, R., Cuninghame-Green, R., Zimmermann, U. (eds) Algebraic and Combinatorial Methods in Operations Research. North-Holland Mathematics Studies, vol. 95, pp. 201–214. North-Holland (1984)
7. Harary, F., Lim, M.-H., Wunsch, D.C.: Signed graphs for portfolio analysis in risk management. IMA J. Manag. Math. **13**(3), 201–210 (2002)
8. Heider, F.: Attitudes and cognitive organization. J. Psychol. **21**, 107–112 (1946)
9. Klamt, S., von Kamp, A.: Computing paths and cycles in biological interaction graphs. BMC Bioinf. **10**(1), 181–192 (2009)
10. Magnanti, T.L., Wolsey, L.A.: Optimal trees. In: Ball, M., Magnanti, T.L., Monma, C., Nemhauser, G.L. (eds.) Network Models, Handbooks in Operations Research and Management Science, vol. 7, pp. 503–615. Elsevier Science Publishers, North-Holland (1995)
11. Wolsey, L.A.: Integer Programming. Wiley, New York (1998)

The Break Point: A Machine Learning Approach to Web Breaks in Paper Mills

Márcia Dias, Nuno Lourenço, Cristóvão Silva, and Samuel Moniz

Abstract Having efficient manufacturing processes requires accurate failure detection to reduce equipment downtime. This paper presents a machine learning approach for predicting web breaks in tissue paper machines. Web breaks prediction plays a key role in ensuring product quality and sustainable use of energy, water, and other resources. The proposed approach can identify moments of high risk of web break occurrence during the regular operation of the paper mill. A large-scale industrial problem from a paper company is used to validate the machine learning model. Results show that web breaks are successfully classified with an accuracy of 86%, reducing production losses by up to 6000 tons of paper per year and cutting water waste by 100.000 litres per day. Indeed, the proposed approach can properly detect operational deviations and alert to a high risk of web breaks, which avoids possible incoming failures and equipment downtimes, helping to secure product quality and a sustainable use of the resources.

Keywords Machine learning · Decision trees · Paper industry · Web breaks

1 Introduction

The production of tissue paper involves two main steps. The first step consists of making the paper pulp, while the second step is responsible for transforming it into paper. This second step is performed in a tissue machine where three main stages

M. Dias (✉) · C. Silva · S. Moniz
Department of Mechanical Engineering, University of Coimbra, Polo II - Pinhal de Marrocos, 3030-290 Coimbra, Portugal
e-mail: marcia.dias@dem.uc.pt

S. Moniz
e-mail: samuel.moniz@dem.uc.pt

N. Lourenço
Department of Informatics Engineering, University of Coimbra, Polo II - Pinhal de Marrocos, 3030-290 Coimbra, Portugal

J. P. Almeida et al. (eds.), *Operational Research*, Springer Proceedings in Mathematics & Statistics 411,
https://doi.org/10.1007/978-3-031-20788-4_5

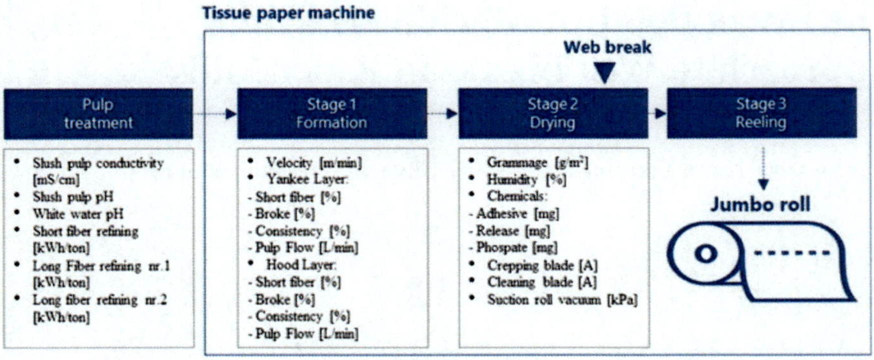

Fig. 1 Process stages in a paper machine and variables selection

occur: formation, drying, and reeling (see Fig. 1). The process starts by compressing the pulp in the formation roll to create the web, which is then forwarded to the drying stage to be dewatered by mechanical and evaporation processes. In the final stage, the web is reeled into the final product—jumbo rolls. During this process, anomalies may result in the breakage of the web. According to the literature, web break problems result in production losses varying between 7 and 12% [3, 4]. However, the waste generated can be even higher, with web breaks representing production losses of up to 35% per day, which is the case we address in this work. In addition to a significant decline in the production output, web breaks also increase the number of defects in the jumbo rolls, which consequently deteriorates the value of the paper in the market. Thus, predicting web breaks contributes not only to improve the productivity, quality, and value of tissue paper, but also to the operation of sustainable production processes.

To address this problem, we propose a data-based procedure to predict, in real-time, the occurrence of web breaks in tissue paper machines. Relevant works focused on predicting web breaks present less integrated models as they use modelling variables determined in an offline manner [12]. Here, we developed a machine learning model that aims to anticipate the occurrence of web breaks with high accuracy and in an online fashion while using a reduced dataset. Although paper machines are usually heavily instrumented with many sensors, missing data often occurs due to the sensors' failures. Thus, fewer modelling variables are used to develop a more robust model to sensor errors. Our work contributes to the literature in three significant ways:

1. tissue paper is a technological product, in which hundreds of variables influence its manufacturing process. Our approach can effectively translate this process into a reduced set of variables and successfully identify web break patterns. Using fewer modeling variables, compared with preceding models from the literature, avoids data unavailability, minimizes the effects of sensor errors on the model, and provides a more compact data-driven procedure. Besides, further information about the process can be provided in real-time to the machine operators that can act preventively before the web breaks occurrence;

2. we propose a data-driven model that uses real-world production data to support effective decisions that leverage the productivity of recent and technologically advanced tissue machines. Even though modern tissue machines have been designed to prevent web breaks and improve efficiency, those are still associated with significant production downtimes;
3. the proposed data-driven procedure copes with the increasing pressure to leverage the production efficiency of continuous production environments.

1.1 Problem Description

The problem under study is to predict web breaks. Web breaks are critical process anomalies that are generally hard to predict due to the difficulty of investigating their root causes. These anomalies tend to occur more frequently at the end of stage 2 of the tissue paper machine, as shown in Fig. 1. Jumbo rolls that exceed the number of breaks that the client is willing to accept are discarded. These rolls are reincorporated in the process, resulting in a loss of capacity and increased production costs.

Compared with previous works in the literature, our approach uses a reduced set of variables and data measurements due to the impossibility of capturing data from some sensors in an accurate and reliable way. Therefore, we did not consider other variables that are typically related to web breaks, such as web tension, web resistance, holes in the web, and side cut quality. And, for instance, we only consider the fiber's refining as a binary variable (showing if it is in use or not in use), instead of being analyzed by its power, pH, or time in use. Moreover, we did not use variables that are measured downstream of the web break. This means that quality measurements, such as paper strength, were not considered. These modeling decisions avoid the use of variables that are not measured automatically by the machine sensors, preventing human errors and favoring an online application of the model. With this procedure, we can overcome data unavailability and minimize the impact of measurement errors on the model, resulting in a more compact approach.

Lastly, there is another additional point that distinguishes our industrial case. Since this study has been conducted in a very recent machine, we can argue that modern and highly technological machines still suffer web break problems. Thus, the development and application of data-driven methods are essential to increase the efficiency of these complex processes.

This paper is organized as follows: Sect. 3 presents the proposed data-driven methodology. Section 4 presents the principal results of the approach and its critical discussion. Finally, Sect. 5 presents the conclusions of the paper.

2 Related Literature

Machine learning methods can be categorized as supervised and unsupervised learning. In supervised learning, the algorithm is provided with a dataset containing the actual outcome for each training example. In contrast, in unsupervised learning, the algorithm only has the input of each training example and needs to model the underlying structure and distribution of the data [1].

The method to use depends on the problem we want to solve: in supervised learning, classification methods try to find classes for a dataset, and regression methods attempt to predict a real value for a given point; in unsupervised learning, clustering methods try to group similar classes and dimensionality reduction methods seek to transform data from a high to a low-dimensional space. Since industrial processes are increasingly equipped with sensors, these methods can effectively reduce operational downtimes and quality problems and improve maintenance procedures [2].

In the last decades, several data mining and machine learning methods have been used in the paper industry to solve production process problems. Bonissone et al. [4] tackle web breaks in stage 1 of paper machines. Alzghoul et al. [12] identify the main parameters affecting the occurrence of web breaks in a paper printing machine, and Alonso et al. [6] predict the properties of the printed paper using machine parameters.

When the cause of the failures is believed to be known, case-based reasoning method can be applied. Ahola et al. [3]use this method and mention the issues of having wrong or incomplete information. Nagappan et al. [7] suggest using a logistic regression for classification when there is a binary response variable. However, this approach is not suitable when the independent variables are highly correlated, and the output variable is not linearly separable, which happens with our case. Musa [8] proposes the utilization of principal components analysis to overcome this issue.

It is also important to mention that the works addressing web breaks in paper mills tend to use a large number of variables that go up to 141. The variables may be related to several parts of the production process: raw material, pulp treatment, paper machine, or paper quality measurements; or related to operational information taken from the shop floor. However, using a large selection of variables does not necessarily mean a better model performance (see [4, 12]). As variables may be taken from both an online and offline approach, it means they are more or less prone to human errors. Overall, the literature shows that decision trees are a good option when there are time limitations and to capture the systems' non-linearities. Also, there is an inherent versatility to these methods, which can be used for classification or regression.

In conclusion, we noted that not many recent works address web breaks' prediction, even though this is still a relevant problem for the paper industry. Hereupon, past models do not consider and do not take advantage of the high level of digitalization of modern paper machines. Also, few works predict web breaks in tissue machines, even though tissue paper production is complex and more sensitive to web breaks.

3 Methodology

Our approach to predict the risk of web breaks comprises five steps as depicted in Fig 2.

Step 1: Variable and Data Selection

The variables selection procedure is based on a qualitative analysis of the production process, firmly supported by the knowledge of the process practitioners. In the first place, only the variables directly linked to the tissue machine are considered. Industrial practitioners also provided useful knowledge to better understand the machine functioning and, therefore, support the variables selection. As listed in the lower part of Fig. 1, 23 independent variables are automatically captured from machine sensors and included in the modeling procedure. In short, the selected data corresponds to two months of machine functioning with a sample interval of 1.5 min. This corresponds to approximately 60,000 observations and a total of 1342 web breaks, which results to an average of 22 web breaks per day.

We note that our approach only uses real data directly taken from the shop floor, meaning that we do not use ideal setups; thus, we consider all the variability that the system is subject to.

Step 2: Data Processing and Segmentation

Data cleaning is performed first and refers to the elimination of contradictory data that do not represent the way the process works. These cases might occur due to low-reliable data measurements and data captured from non-operating periods of time. In our case, some sensors keep falsely measuring and registering values even when the machine is idle. Furthermore, data just after a web break occurs is removed since it does not represent the typical operation of the machine. At last, we also remove all the well-known breaks so as to analyze only the unknown web break trajectories and to reduce the computational complexity.

Subsequently, since it is not possible to identify all failures in a reliable way, we use data segmentation (see [4]) to divide the data per web break trajectory, considering a minimum time interval of 3 h of machine functioning until the web break occurrence. With this approach, the proposed model will use a tidy dataset and clean web break signals that significantly improve the quality of the data and help to avoid overfitting.

Like Ahola et al. [3], we defined the web break risk within each web break trajectory as follows: high, medium, low and null risks—depending on the time to failure. As shown in Fig. 3, we have considered one-hour intervals for high, medium, and low risks and the remaining period is assumed to have a null risk. This concept of web break risk provides valuable information to the machine's operators for preventing the occurrence of web breaks. In practice, the risk levels work as a warning system in our prediction model. If the returned value is null or low, operators perceive that the system is working well. If the risk is medium, they must be alert to what was changed and adjust values accordingly. And when the risk is high, they know that there may be an imminent break.

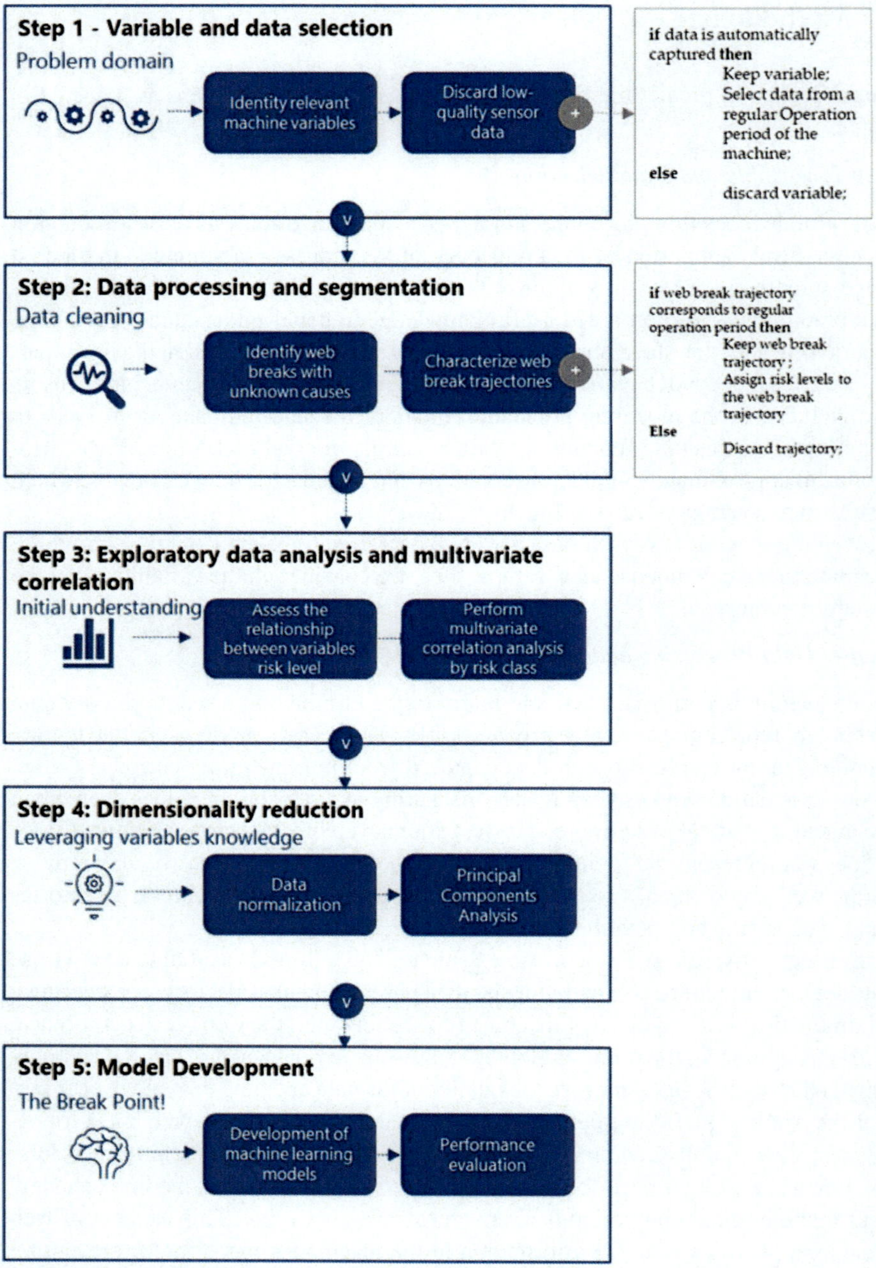

Fig. 2 Flowchart of the proposed approach

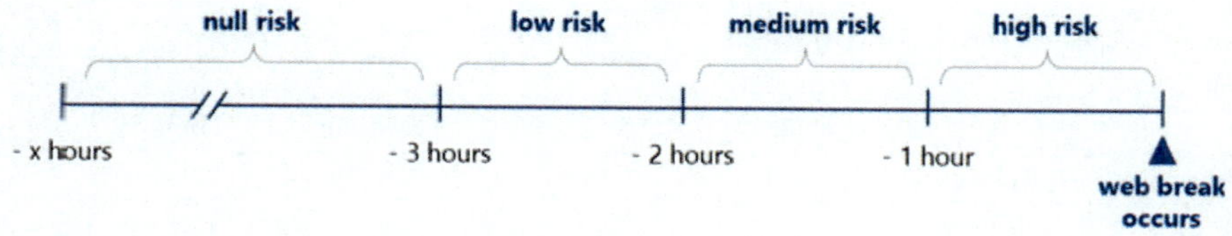

Fig. 3 Risk levels in a web break trajectory

Step 3: Exploratory Data Analysis and Multivariate Correlation

At this point, an exploratory data analysis is done to perceive some relations between web breaks and variables. We use multivariate correlation analysis by risk class to find if there are any changes in the relationships between variables depending on the risk level. We used Pearson's coefficient [10] for both these analyses, and we resorted to the scikit-learn library in python programming language to do so [11].

Step 4: Dimensionality Reduction

Dimensionality reduction using a principal components analysis is suggested since it can lead to more simple models. This method intends to transform the dataset into a lower dimension without losing significant or valuable information, such as trends and patterns.

Step 5: Model Development

Reliably improving the accuracy for web breaks' risk prediction is achievable by training distinct classification and regression trees (CART) models. To build a decision tree classifier, we divided the normalized dataset into training (70%) and testing (30%) sets. Then, we applied a Decision Tree classifier with the Gini index metric as the impurity metric to the train set. Finally, we evaluated the generalization ability of the classifier on the test set. As performance metrics, we considered accuracy, precision, and recall. Beyond that, we also resorted to the confusion matrix to verify if there is a pattern of confusion while predicting specific risk levels.

4 Results and Discussion

In this section, we provide the implementation details of the proposed approach.

Step 1: Variable and Data Selection

The machine produces tissue paper following three main stages, starting from the pulp to jumbo rolls of tissue paper. A total of 23 independent variables, 60,000 observations, and 1342 web breaks were automatically captured from machine sensors and included in our methodological approach.

Table 1 Pearson's correlation coefficients

Variables		Risk level			
		Null	Low	Medium	High
Adhesive	Release	0.8	0.9	0.9	0.9
YL flow	HL flow	1	1	1	1
Adhesive	Velocity	−0.7	−0.6	−0.7	−0.7
Slush conductivity	Slush pulp	0.6	0.5	0.5	0.5
Suction roll vacuum	Velocity	−0.5	−0.5	−0.6	−0.6
Grammage	Velocity	−0.2	−0.2	−0.2	−0.2

Step 2: Data Processing and Segmentation

After data cleaning, the total number of observations was reduced by 23% and the number of web breaks decreased 26%. Moreover, due to data segmentation, we noted a decrease of 58% of the observations and a decrease of 89% in the number of web breaks signals.

Step 3: Exploratory Data Analysis and Multivariate Correlation

In the exploratory data analysis we observed that 2 out of the 23 variables are constant over time. These variables have been discarded since they have no explanatory power of web breaks. Henceforward, only 21 independent variables were considered. We used a multivariate correlation analysis to look for changes in the relationships between variables as a function of the risk level. Table 1 presents the most relevant results.

As expected, there is a strong positive correlation between the adhesive and release chemicals, which is transversal to all risk levels. Having a higher adhesive chemical usage means a higher adherence of the paper to the heated cylinder, and therefore it is necessary to use a large amount of release chemical in order to launch the paper from its surface. The same goes for the yankee and hood layer flows. There is also a strong negative correlation between the adhesive chemical and the machine's velocity across all risk levels. On the other hand, a decrease in the correlation is perceived between slush conductivity and slush pH as the risk level increases. Also note that there is a strong negative correlation between suction roll vacuum and machine's velocity, which demonstrates the need for a lower pressure in the suction roll vacuum when the machine's velocity is higher.

Step 4: Dimensionality Reduction

Principal Components Analysis was used to reduce the dimensionality of the problem. Figure 4 shows the percentage of variance explained by each principal component and its accumulated variance, and Fig. 5 depicts the contributions of the variables to each principal component.

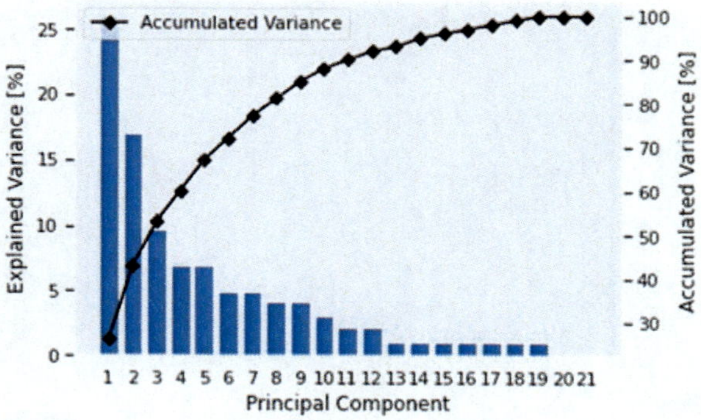

Fig. 4 Percentage of variance explained by each principal component

Table 2 Confusion matrix of the model with 4 risk levels (%)

Predicted risk	Real risk			
	Null	Low	Medium	High
Null	96	5	0	1
Low	2	85	12	8
Medium	1	7	78	10
High	1	3	10	81

We observed that grammage, humidity, short fiber refining, and the percentages of short fiber and broke used on the yankee layer have a more significant contribution. On the other hand, variables such as release and adhesive chemicals, yankee layer flow, suction vacuum, and slush pH are less relevant. Note also that it is perceived a prevalence of variables belonging to stage 1 over stage 2.

The first three components showed a prevalence of humidity and the percentages of short fiber and broke used on the yankee layer. Figure 5 also reveals that, even though slush pH has a low contribution to the first component, it has a strong presence from the second to the sixth component. Something similar happens with the suction roll vacuum, which presents high contributions in the second and fourth components. In this paper, in order to decide how many principal components to keep, a combination of three techniques is suggested: (i) eigenvalue greater than 1 (Kaiser–Guttman criteria); (ii) cumulative variance greater than 90%; and (iii) observation of the eigenvalues graph. According to these criteria, twelve principal components should be selected to further use. However, we considered it is not beneficial to use the principal components, as the overriding objective of this step was to significantly reduce the dimensionality of the problem.

Step 5: Model Development

In the last step of our approach, we used classification and regression trees (CART) to develop the prediction model. We analyzed the accuracy of the model using the

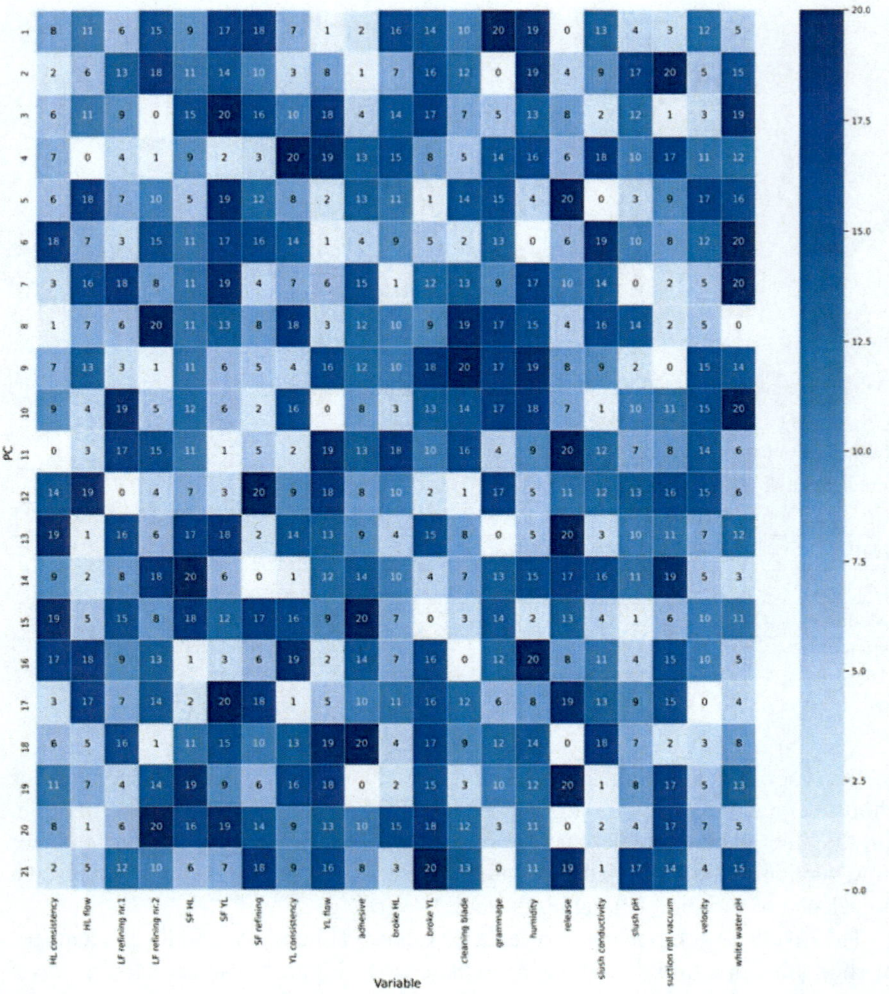

Fig. 5 Contributions of the variables to each principal component

test dataset and varying the maximum depth of the decision tree. The aim was to find the trade-off between accuracy gain and loss of interpretability, as more depth levels lead to a more complex tree. Results shown in Fig. 6 demonstrate that using 21 depth levels, the model achieves an accuracy of 86%.

The confusion matrix presented in Table 2 shows that the model correctly predicted the null risk 96% of the times, the low risk 85%, the medium risk 78%, and the high risk 81%. We can conclude the model has difficulty discriminating between medium and high risks. While predicting medium level risks, more than 20% of the predictions are false low or high. Also, there is difficulty predicting high risk, as the

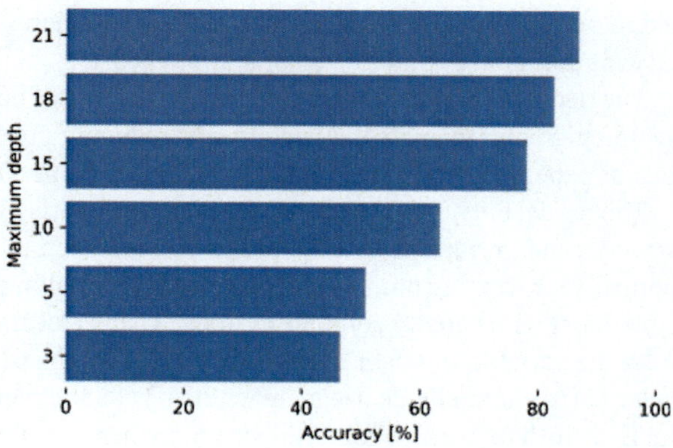

Fig. 5 Prediction accuracy according to the maximum depth of the decision tree

Table 3 Precision and recall of the model with 4 risk levels and 21 variables (%)

Predicted risk	Precision	Recall
Null	97	96
Low	78	85
Medium	80	78
High	86	81

model wrongly returns medium risks 10% of the times. However, overall, we can say that the model can discriminate between different risk levels.

In Table 3, we present the model's precision and recall. It can be seen that the model successfully predicts the null risk, presenting a precision of 97% and recall of 96%. In our problem, precision plays an important role, as a false prediction of high risk of a web break would not be prejudicial to the production process. However, some importance should be given to recall, as it is also beneficial to predict every high risk of a web break. Thus, even though the high-risk level has a precision of 86%, its recall goes down to 81%, meaning that almost 20% of the times the model did not correctly alert to a high-risk web break.

5 Conclusions

In the pulp and paper industry, predicting and avoiding web breaks means significant gains in productivity and efficiency. However, making a solid prediction is hampered by the manufacturing complexity and the consequent difficulty in identifying patterns. In this work, we develop a prediction model for the risk of web break occurrences. Even though modern tissue machines are highly technological, web breaks are still the cause of relevant production losses in the paper industry. Classification and regression trees were the chosen method for addressing the case study, given its easy interpretability and low computational cost.

The decision-tree model predicted the risk level of occurring a web break with an accuracy greater than 86%, which means reducing production losses by up to 6000 tons of paper per year and a reduction of more than 100.000 liters of water per day.

The model only uses variables related to three parts of the process: pulp treatment, wet-end, and drying stage of the tissue machine; and all of them are measured automatically, preventing human-induced errors and making it easier to replicate to other shop floors with equally complex systems. Using a decision tree model allowed us to order the variables by their decision power. It is suggested special attention to white water pH, cleaning blade usage, machine's velocity, paper grammage, broke used in the HL, slush pH, and suction roll vacuum when faced with a high risk of web break occurrence. Furthermore, we emphasize that a regular usage of the cleaning blade is a good option to prevent failures.

Lastly, the proposed model is suitable to be part of a more complex analysis, where it addresses not only single machine failures but also factory interdependencies: i.e., it can be replicated to other process machines, which are then combined with each other, offering a large-scale view of a given factory. Such results can be used to prevent delivery delays and aid production planning decisions.

This work presents itself as a foundation for a more robust and complete model. Thus, as future works, it is suggested to use variables that are considered relevant but were not available for this analysis. Furthermore, it is also important to take into consideration the time dependencies between the observations. This last suggestion seeks to provide answers about the time that the machine may take before changing the risk level.

Acknowledgements This research was funded by FCT—Fundação para a Ciência e a Tecnologia, grant number UID/EMS/00285/2020; and by the Operational Programme for Competitivity and Internationalization of Portugal 2020 Partnership Agreement (PRODUTECH4S&C), grant number POCI-01-0247-FEDER-046102; and by project CISUC—UID/CEC/00326/2020 and by European Social Fund, through the Regional Operational Program Centro 2020.

References

1. Xiao, J., Tian, Y., Xie, L.: A hybrid classification framework based on clustering. IEEE Trans. Ind. Inf. **16**, 2177–2188 (2020). https://doi.org/10.1109/TII.2019.2933675
2. Wang, J., Ma, Y., Zhang, L.: Deep learning for smart manufacturing: methods and applications. J. Manuf. Syst. **48**, 144–156 (2018). https://doi.org/10.1016/j.jmsy.2018.01.003
3. Ahola,T., Kumpula, H., Juuso, E.: Prediction of paper machine runnability by identification of operating situations. IFAC Proc. Vol. **37** (2004). https://doi.org/10.1016/S1474-6670(17)30875-3
4. Bonissone, P., Goebel, K., Chen, Y.T.: Predicting wet-end web breakage in paper mills. In: Working Notes of the 2002 AAAI Symposium: Information Refinement and Revision for Decision Making: Modeling for Diagnostics, Prognostics, and Prediction, pp. 84–92 (2002)
5. Alzghoul, A., Verikas, A., Hållander, M., Bacauskiene, M., Gelzinis, A.: Screening paper runnability in a web-offset pressroom by data mining. In: Perner P. (eds.) Advances in Data

Mining. Applications and Theoretical Aspects. ICDM 2009. Lecture Notes in Computer Science, vol. 5633. Springer, Berlin (2009). https://doi.org/10.1007/978-3-642-03067-3_14
6. Alonso, A., Negro, C., Blanco, A., San Pío, I.: Application of advanced data treatment to predict paper properties. Math. Comput. Modell. Dyn. Syst. **15**(5), 453–462 (2009). https://doi.org/10.1080/13873950903375445
7. Nagappan, N., Ball, T., Zeller, A.: Mining metrics to predict component failures. In: Proceedings - International Conference on Software Engineering 2006, pp. 452–461 (2006). https://doi.org/10.1145/1134285.1134349
8. Musa, A.B.: A comparison of $\ell 1$-regularizion, PCA, KPCA and ICA for dimensionality reduction in logistic regression. Int. J. Mach. Learn. & Cyber. **5**, 861–873 (2014). https://doi.org/10.1007/s13042-013-0171-7
9. Niskanen, K.: Mechanics of Paper Products. Walter de Gruyter, Berlin (2011)
10. Ly, A., Marsman, M., Wagenmakers, E.J.: Analytic posteriors for Pearson's correlation coefficient. Statistica Neerlandica **72**, 4–13 (2018). https://doi.org/10.1111/stan.12111
11. Hao, J., Ho, T.K.: Machine learning made easy: a review of Scikit-learn package in python programming language. J. Educ. Behav. Stat. **44**, 348–361 (2019)
12. Alzghoul, A., Verikas, A., Hållander, M., Bacauskiene, M., Gelzinis, A.: Screening Paper Runnability in a Web-Offset Pressroom by Data Mining. Lecture Notes in Computer Science, vol. 5633. Springer, Berlin (2009). https://doi.org/10.1007/978-3-642-03067-3_14

A Resectorization of Fire Brigades in the North of Portugal

Maria Margarida Lima, Filipe Soares de Sousa, Elif Göksu Öztürk, Pedro Filipe Rocha, Ana Maria Rodrigues, José Soeiro Ferreira, Ana Catarina Nunes, Isabel Cristina Lopes, and Cristina Teles Oliveira

Abstract Sectorization consists of grouping the basic units of a large territory to deal with a complex problem involving different criteria. Resectorization rearranges a current sectorization avoiding substantial changes, given a set of conditions. The paper considers the case of the distribution of geographic areas of fire brigades in the north of Portugal so that they can protect and rescue the population surrounding the fire stations. Starting from a current sectorization, assuming the geographic and population characteristics of the areas and the fire brigades' response capacity, we provide an optimized resectorization considering two objectives: to reduce the rescue time by maximizing the compactness criterion, and to avoid overload situations by maximizing the equilibrium criterion. The solution method is based on the Non-dominated Sorting Genetic Algorithm (NSGA-II). Finally, computational results are presented and discussed.

Keywords Sectorization · Resectorization · Fire brigades · Multi-objective problems

M. M. Lima (✉) · A. M. Rodrigues · I. C. Lopes · C. T. Oliveira
CEOS.PP, ISCAP, Polytechnic of Porto, Porto, Portugal
e-mail: mmal@iscap.ipp.pt

F. S. de Sousa · A. C. Nunes
ISCTE, University Institute of Lisbon, Lisbon, Portugal

E. G. Öztürk · P. F. Rocha · A. M. Rodrigues · J. S. Ferreira
INESCTEC, Technology and Science, Porto, Portugal

J. S. Ferreira
FEUP, Faculty of Engineering, University of Porto, Porto, Portugal

E. G. Öztürk
FEP.UP, Faculty of Economics, University of Porto, Porto, Portugal

A. C. Nunes
CMAFcIO, Faculty of Sciences, University of Lisbon, Lisbon, Portugal

J. P. Almeida et al. (eds.), *Operational Research*, Springer Proceedings in Mathematics & Statistics 411,
https://doi.org/10.1007/978-3-031-20788-4_6

1 Introduction

In Portugal, fire brigades are the core of civil protection. They have several missions, such as extinguishing fires and pre-hospital emergencies. Prompt response to emergency incidents is primordial since delays in the rescue can have tragic consequences. However, increasing urban development presents an increase in demand and a challenge to the emergency response mechanisms. Fire brigades must protect and rescue the population in the areas surrounding their fire stations. This paper intends to contribute, to this desideratum, by applying the concept of sectorization to a real case.

Given a large territory or network composed of basic units or indivisible regions, a sectorization consists of grouping the basic units into sectors considering one or more objectives, intending to simplify a problem. Resectorization intends to change a current sectorization by avoiding substantial changes, but responding to a set of conditions. The term resectorization is also known as redistricting, as indicated in the references [5, 11].

Resectorization will be applied to the current sectorization of a real situation in northern Portugal. The goal is to create compact and balanced sectors of the fire brigades' operation to decrease rescue time and avoid work overload. For that, we also propose new measures to calculate the compactness and equilibrium of the sectorization. To solve the fire brigades problem, we will employ a popular multi-objective optimization method named Non-dominated Sorting Genetic Algorithm (NSGA-II) [6]. This algorithm has also been used to solve sectorization problems [7, 8].

The remainder of the paper is organized as follows. Section 2 presents a review of the relevant literature. Section 3 describes the case study and also the data used. Section 4 explains the solution method. Section 5 presents the results and a brief discussion. Finally, Sect. 6 puts forward the conclusions.

2 Relevant Literature

This section briefly summarizes some of the literature relevant to sectorization problems.

Fire brigades have been the subject of several studies over time. One of the topics covered is to understand the characteristics and incidence of fires to improve the rescue time and reduce their consequences [4, 12].

Sectorization problems have many real-world applications. Political districting is one of the oldest, aiming to divide the territory neutrally and avoid gerrymandering. Bozkaya et al. [3] propose a tabu search and adaptive memory heuristic to solve a single multicriteria problem involving contiguity, population equality, compactness and socio-economic homogeneity, using real data from Edmonton. Bação et al. [1] use a genetic algorithm to solve the problem of electoral districting.

Regarding resectorization, political redistricting is a common application [9]. It intends to redesign the boundaries of existing legislative districts for electoral purposes. Using a contiguity procedure, one of the used algorithms exchanges population units between a district with a population smaller than the ideal and a district with a larger population.

Moreover, Assis et al. [5] applied resectorization to meter reading in power distribution networks and proposed a Greedy Randomized Adaptive Search Procedure (GRASP) to solve a case with real-world customers in Brazil. Vahdani et al. [11] proposed a bi-objective optimization model to plan a humanitarian districted logistics network, with several simultaneous decisions concerning emergency facility location-allocation, redistricting and service sharing.

Another field studied is school redistricting, which consists of adjusting the boundaries of schools [2]. In this paper, the authors assign students to schools considering a balance of the schools' socioeconomic compositions and the total travel distance.

3 Case Study

This section describes the resectorization of fire brigades case study. It involves 6 fire brigades and the 175 indivisible regions they serve, situated in the north of Portugal.

Let us define the following parameters:

$R = \{1, ..., n\}$ the set of indivisible regions;
$B = \{1, ..., m\}$ the set of fire brigades;
d_{ji}, the euclidean distance between the center of region $j \in R$ and the brigade $i \in B$;
p_j, the number of inhabitants in region $j \in R$;
q_j, the area of the region $j \in R$;
A_i, the number of ambulances of the brigade $i \in B$;
V_i, the number of fighting vehicles of the brigade $i \in B$;
e_i, the number of firefighters of the brigade $i \in B$;

$$X_{ji}^E = \begin{cases} 1, \text{ if region } j \text{ is assigned to brigade } i \text{ in the current sectorization} \\ 0, \text{ otherwise} \end{cases}$$

The demand of region s_j depends on its number of inhabitants and its area, taking the form (Fig. 1)

$$s_j = \alpha_1 P_j + \alpha_2 Q_j, \tag{1}$$

where α_1 and α_2 are constants,

$$P_j = \frac{p_j}{\sum_{j \in R} p_j} \tag{2}$$

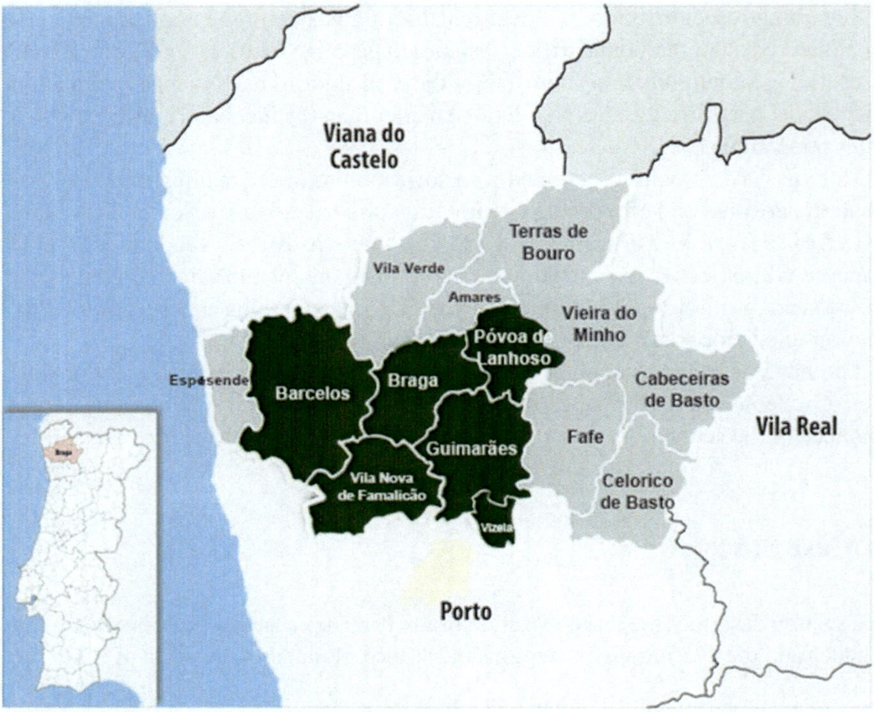

Fig. 1 The territory addressed in this study is represented in green, located in the north of Portugal

and

$$Q_j = \frac{q_j}{\sum_{j \in R} q_j}. \tag{3}$$

As the regions have a large population diversity and areas, these values had to be adimensionalized. The proportion of each region to the total was considered to join both values in the same formula.

These two characteristics of the regions, P_j and Q_j, are used to categorize the data better. The constants α_1 and α_2 will be used as relative weights of these two characteristics, adjusted according to specific needs. These values are related to the probability of a rescue event which depends differently on the population and the area of the region. It is assumed that the larger the population, the greater the probability of being necessary ambulances and firefighters. Also, a region with a larger area will probably have more incidents, requiring more vehicles and firefighters.

The capacity of each fire brigade i derives from its number of ambulances, its number of fighting vehicles and its number of firefighters, and is given by

$$c_i = \beta_1 g_1 A_i + \beta_2 g_2 V_i + \beta_3 E_i, \tag{4}$$

where

$$E_i = \frac{e_i}{\sum_{i \in B} e_i}. \tag{5}$$

Constant parameters g_1 and g_2 have been adjusted to reflect the current state of the most overloaded fire brigade in set B. g_1 is the ratio between the total population currently served by this corporation and its number of ambulances. Similarly, g_2 is the ratio between the total area currently served by this corporation and its number of fighting vehicles.

The constants β_1, β_2 and β_3 represent the weights of each characteristic of the fire brigades, which can be adjusted based on the population's needs.

3.1 *Objective Functions*

The case study aims to create sectors where service centers are the fire stations and the indivisible regions are the basic units. For this, two objective functions are chosen, related to the criteria: Equilibrium and Compactness.

Let us define the following variables:

$$x_{ji} = \begin{cases} 1, \text{if region } j \text{ is assigned to fire brigade } i \text{ in the resectorization} \\ 0, \text{otherwise} \end{cases}$$

Equilibrium:

The idea is to distribute the regions equitably among the fire brigades, being necessary to assume the capacity of each fire brigade, c_i, and the demand of each region, s_j.

The objective is to minimize the standard deviation of the occupancy percentage of fire brigades:

$$\min f_1 = \sqrt{\frac{\sum_{i \in B}(k_i - \bar{k})^2}{m}}, \tag{6}$$

where $k_i = \frac{\sum_{j \in R} s_j x_{ji}}{c_i}$ is the percentage of used capacity of fire brigade i and $\bar{k}$ is the average of k_i.

This objective function was created to obtain an equitable distribution of demands. It is necessary to consider each corporation's capacity to minimize the possibility of overload.

Compactness:

The view is to distribute the regions among the fire brigades such that the rescue time is minimal. For this, we attend to the distance between the fire departments and the regions, d_{ji}, and the regions' needs, s_j.

The objective is to minimize the distance between fire departments and regions, weighted by the demand of each region:

$$\min f_2 = \sum_{j \in R} \sum_{i \in B} d_{ji} s_j x_{ji}. \tag{7}$$

By minimizing the distance between regions and the assigned corporation, we expect to minimize the rescue time. Additionally, the demand is regarded since a region that needs to be visited more frequently will have greater weight because it will take more time.

3.2 *Similarity of the Solutions*

The similarity is a measure to evaluate the resemblance, or coincidence, between two solutions. In resectorization, it is important to measure how different is the new solution from the original one. The importance of this measure arises clearly in firefighters' cases to facilitate the adaptation of firefighters to changes.

The similarity measure used in this work is the percentage of regions that stay unchanged, which is given by:

$$\frac{\sum_{j \in R} \sum_{i \in B} X_{ji}^E x_{ji}}{|R|} \geq \delta, \qquad \delta \in [0, 1]. \tag{8}$$

This measure is used as a constraint. The decision-maker will define the preferred minimum similarity level, δ.

4 Solution Method

The paper follows NSGA-II to resectorize the fire brigades. NSGA-II is one of the most recognized and implemented multi-objective optimization (MOO) methods. As in all MOO methods, NSGA-II classifies all the solutions for each objective simultaneously and separately regarding their performance. These ranks are constructed considering two parameters: (i) set of dominating members, and (ii) domination count. The former counts how many other solutions a solution dominates. The latter counts by how many other solutions a solution is dominated. A solution (say x) dominates another solution (say y) if solution x performs better than solution y for at least one objective while it is not worse than y in any other objectives. This definition helps to build the two sets for each solution. More precisely, the set of dominating members of a solution includes the solutions dominated by that solution. Moreover, the domination count of a solution sums up the number of solutions that dominate that solution. These two sets are constructed for each solution. The solutions with zero domination count are allocated in the first Pareto frontier and considered the best. The solutions that are dominated by the solutions located in the first Pareto frontier are allocated in the second Pareto frontier. This sequence continues until all the solutions are allocated in the Pareto frontiers. The solutions in the same Pareto frontier are regarded as equally good. This domination logic ultimately helps keep the best solutions concerning their performances over generations in the objective space.

As in all evolutionary algorithms, NSGA-II starts by initializing the population. The population usually includes a set of solutions that are randomly generated. In the case of resectorization, the solutions must follow a specific order based on the existing solution (i.e. current sectorization) while still allowing for some randomness.

In this work, the "matrix form binary grouping" (MFBG) genetic encoding system [10], suitable for sectorization problems, is used to compose the solutions. MFBG is a binary matrix of size composed of (J × I), where J and I represent the total number of basic units and sectors, respectively. Each row can be considered a binary set representing the sector assignment of the corresponding basic unit.

It is possible to see its format below:

$$M = [x_{ji}]_{J \times I}$$

where

$$x_{ji} = \begin{cases} 1 & \text{if basic unit } j \text{ is assigned to sector } i \\ 0 & \text{otherwise} \end{cases}$$

The solutions included in the population are evaluated according to their performance on the two objectives specified in Sect. 3.1 according to their Pareto dominance in the objective space.

NSGA-II seeks the objective space over generations to find superior solutions. An entire generation (i.e. iteration) consists of some steps, namely, selection, crossover, and mutation, until the stopping criterion occurs.

Selection picks the parent solutions necessary for the crossover. We used tournament selection. In this method, two solutions are randomly selected from the population and evaluated according to their performances for each objective. If there is domination between the solutions, we keep the dominant solution as a parent. If the two solutions are in the same Pareto frontier, since they are not comparable, we observed an NSGA-II specific concept, called Crowding Distance (CD), to select the parent solution. CD represents the concentration of the solutions in the Pareto frontier, such that a solution with a lower value of CD is more crowded with other solutions. Therefore, solutions with higher CD are assumed to be better and more representative of the population to increase diversity.

The selected parent solutions are mated to create off-spring solutions during the crossover. In the current work, we used multi-point crossover, which selects random rows from the parent solutions and switches them to form two off-spring solutions. This crossover method does not require a crossover probability due to the design of the MFBG genetic encoding system.

Finally, the mutation operator is just implemented on the off-springs in a given probability to increase the diversity in the population. The mutation occurs with a certain probability imposed on the population. The algorithm decides row by row if a mutation happens, i.e., if a point will be assigned to another sector or not.

All the solutions are evaluated regarding their similarity to the existing solution using the equation presented in Sect. 3.2 The solutions with a higher difference than the desired similarity level are eliminated from the population. Then, new ones are generated for each deleted solution to keep the population size constant over generations.

The process is described in the Algorithm.

The parameters, namely population size and mutation probability were 100 and 0.03, respectively. Finally, the number of generations was 500 and used as a stopping criterion.

Algorithm 1 Pseudocode of NSGA-II

Generate N feasible solutions and insert into Population ($Pop_{size} == N$)
Evaluation of the solutions according to the selected objectives
Non-Dominated Sorting $MinF(.)$
Calculate Crowding Distance of each frontier
$Generation := 0$
while $Generation < T$ **do**
  while $Pop_{size} < N \times 2$ **do**
    Select parents through tournament selection
    Create two off-springs using multi-point crossover in each turn
    Mutate off-springs (for selected P_{mut})
    Merge off-springs into the population
    $Pop_{size} := Pop_{size} + 2$
  end while
  for each solution in the current population **do**
    Evaluation of the solutions according to the selected objectives
    Non-Dominated Sorting $MinF(.)$
    Calculate Crowding Distance of each frontier
  end for
  while $Pop_{size} \neq N$ **do**
    if $N \geq Pop_{size}$ - # solutions in the worst Pareto frontier **then**
      Remove all solutions located in the worst Pareto frontier.
      $Pop_{size} := Pop_{size}$ - # solutions in the worst Pareto frontier
    else
      Remove the worst solution in the worst Pareto frontier regarding the Crowding Distance
      $Pop_{size} := Pop_{size} - 1$
    end if
  end while
  $Generation := Generation + 1$
  for each solution in the population **do**
    Evaluation of the solutions according to their similarity
    Delete the solutions with higher differences than the desired similarity level
    Generate feasible solutions and insert into Population to keep $Pop_{size} == N$
  end for
  for each solution in the population **do**
    Evaluation of the solutions according to the selected objectives
    Non-Dominated Sorting $MinF(.)$
    Calculate Crowding Distance of each frontier
  end for
  Output the results
end while

5 Results and Discussion

In this section, we will present the results and a brief discussion.

Based on experience, the constants presented in expressions (1) and (4) were fixed.

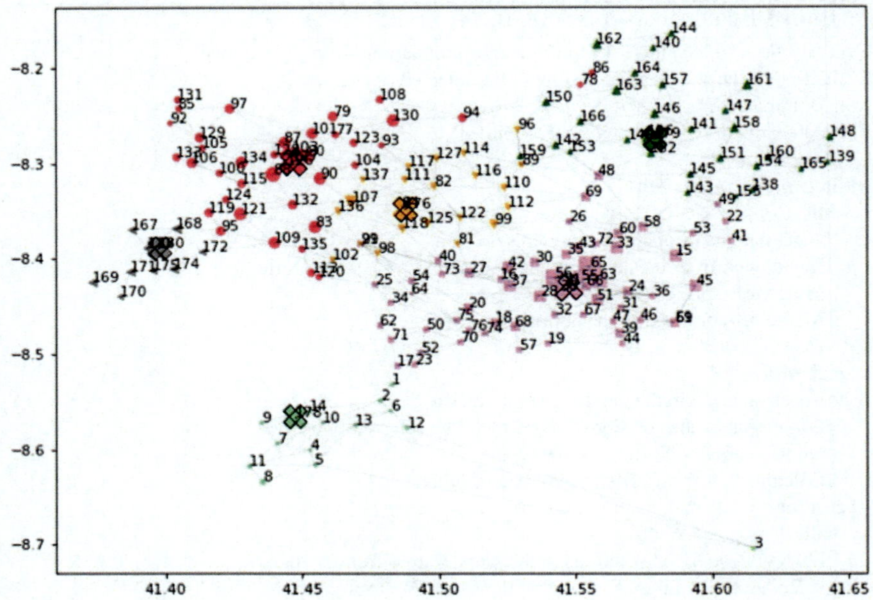

Fig. 2 Current sectorization (Equilibrium = 0.284; Compactness = 0.0035)

Population growth will increase the need for pre-hospital emergencies and, growth in the area of the region may increase the number of forest fires, accidents, etc. The parameters $\alpha_1 = 0.6$ and $\alpha_2 = 0.4$ were respectively defined in expression (1).

It is reported that most of the services provided by firefighters are pre-hospital emergencies. In expression (4), the relative importance of the number of ambulances, the number of other vehicles and the number of firefighters were defined, respectively, $\beta_1 = 0.5$, $\beta_2 = 0.3$ and $\beta_3 = 0.2$.

In Fig. 2, it is possible to see the current sectorization used by the fire brigades. The crosses correspond to the fire brigades stations and the points represent the center of the regions. A different color denotes each sector.

After solving the problem described in Sect. 3, we obtained several solutions, using the similarity as a restriction, taking 70%.

Figure 3 pictures the obtained Pareto Frontier. Equilibrium versus Compactness, in all solutions, with different levels of similarity, are also represented.

We can see ten solutions for optimized solutions with a similarity between 70% and 80%, five solutions with a similarity between 80% and 90% and five solutions with a similarity between 90% and 100%.

Pareto Frontier permits verifying that the best solutions in terms of equilibrium are presented with a similarity between 70% and 80%. On the other hand, better solutions in compactness are obtained with a similarity between 90% and 100%. In addition, it is possible to see that a small improvement in compactness leads to a large deterioration of the equilibrium values.

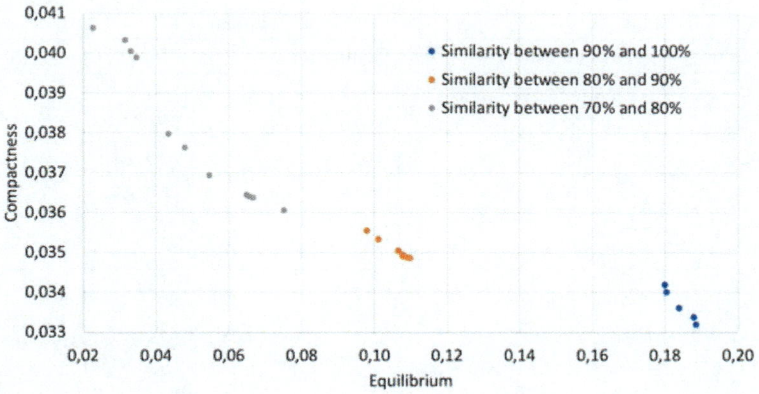

Fig. 3 Pareto Frontier for $\delta = 0.70$

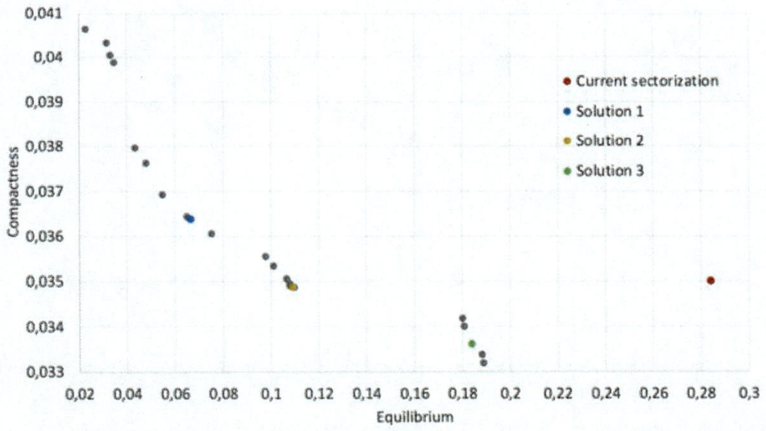

Fig. 4 Current sectorization and optimized solutions for $\delta = 0.70$

It is also possible to see that the equilibrium decreases even for high similarity values (between 70% and 100%). Changing some points to a different sector reduced the equilibrium value substantially. In practice, the regions will be distributed equitably between the fire brigades, reducing the chance of overload.

In order to visualize the improvements, in Fig. 4, we present the equilibrium and compactness values of the current sectorization and the solutions in the Pareto frontier. The solutions in different colors represent the values of objective functions in current sectorization and three optimized solutions with different similarity values that could be suitable (Table 1). All solutions were presented to a person responsible for one of the fire brigades, and these three solutions were pointed out as the most suitable for implementation in the field.

In Figs. 5, 6 and 7, the three proposed resectorizations are shown.

Table 1 Objective functions of the proposed resectorization solutions

Sectorization	Equilibrium	Compactness	Similarity
Current	0.284	0.035	–
Solution 1	0.067	0.036	74.3%
Solution 2	0.109	0.035	80.6%
Solution 3	0.184	0.034	90.3%

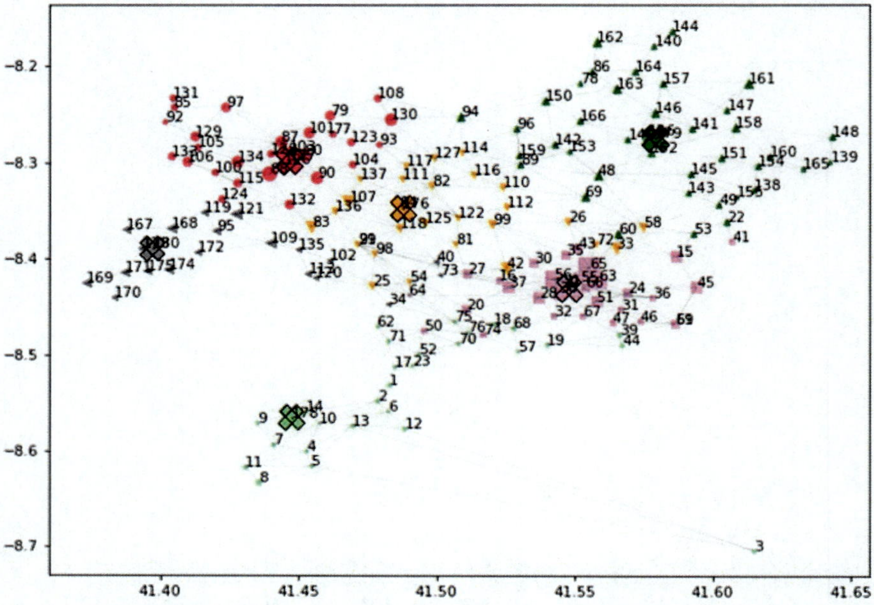

Fig. 5 Solution 1 (Equilibrium = 0.067; Compactness = 0.036; Similarity = 74.3%)

First of all, the solutions presented have better values in terms of equilibrium when compared to the current sectorization, and the lowest equilibrium value is obtained for similarity of 74.3%.

Furthermore, all solutions presented compactness very similar to the current sectorization. However, we can observe a trade-off between compactness and equilibrium, which means better compactness values imply the worst equilibrium values, which follows directly from the concept of nondominated solutions.

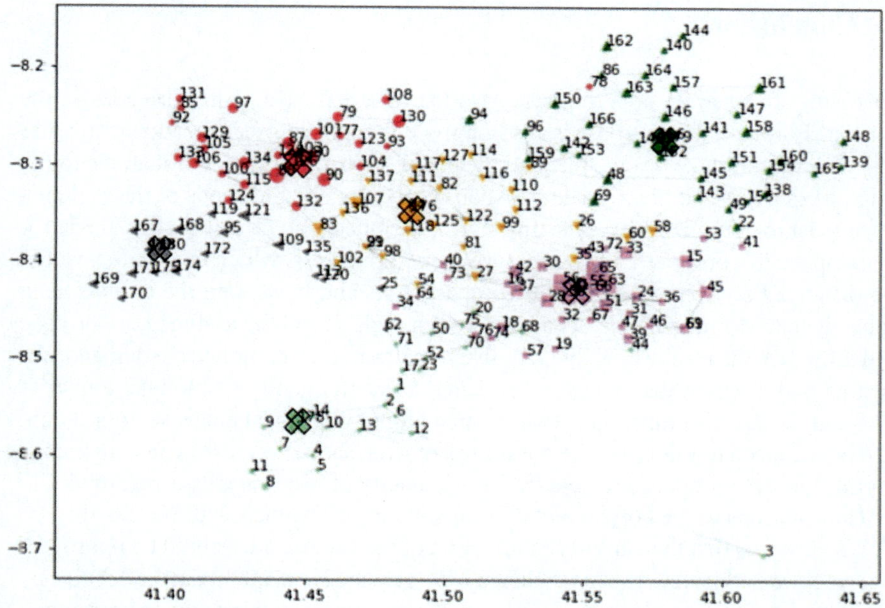

Fig. 6 Solution 2 (Equilibrium = 0.109; Compactness = 0.035; Similarity = 80.6%)

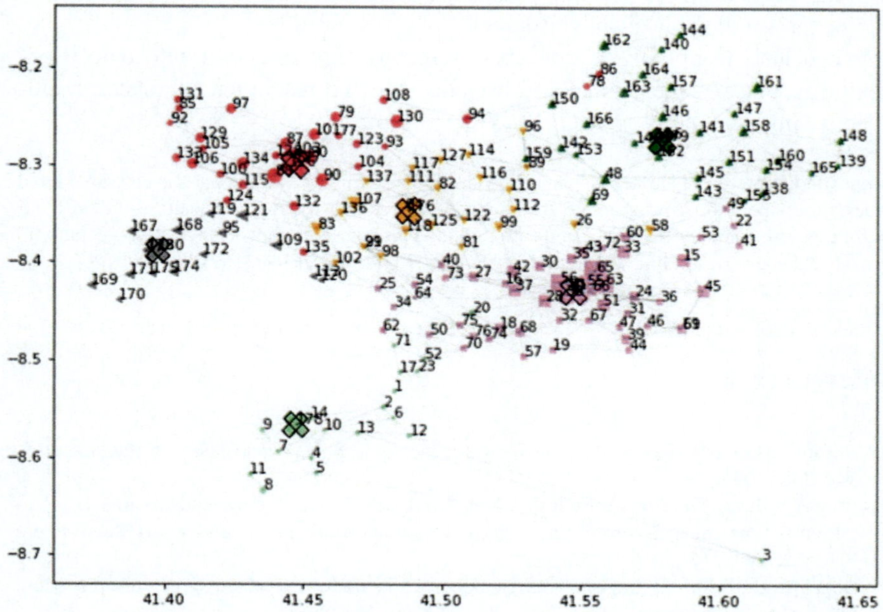

Fig. 7 Solution 3 (Equilibrium = 0.184; Compactness = 0.034; Similarity = 90.3%)

6 Conclusions

This study focused on applying sectorization to a real case of fire brigades in the north of Portugal. The purpose was to improve the existing sectorization of the fire brigades, to minimize rescue time and the possibility of overload. For that, we found a way to quantify the fire brigades' response capacity and the needs of the regions.

In order to minimize the rescue time, it was assumed that the new sectors needed to be compact. We proposed a new measure for compactness, which takes into account the distance between regions and fire departments and how often the regions need to be visited. Minimizing the distance between regions and the assigned fire brigade will minimize the rescue time, but it is also important to contemplate the demand. We assume that a region that needs to be visited more frequently would take longer to save and therefore to this region was given a higher weight. The new sectors should also be balanced to minimize the possibility of overload. Therefore, a new metric for equilibrium was proposed, where the used capacity of fire brigades is regarded.

The solutions to the corresponding problems were obtained with NSGA-II.

We observed that the current sectorization of the fire brigades could be improved; the sectors are compacted but not balanced. We verified an inequality in the occupancy rates of the different fire brigades. We have shown that maintaining good compactness can greatly improve the equilibrium between corporations, minimizing the risk of overload. Besides, three optimized solutions were proposed to reduce rescue time with better use of the available resources.

In conclusion, applying the concept of resectorization can contribute to decision-making in the management of fire brigades and the resolution of related multi-objective problems.

Acknowledgements This work is financed by the ERDF—European Regional Development Fund through the Operational Programme for Competitiveness and Internationalisation—COMPETE 2020 Programme and by National Funds through the Portuguese funding agency, FCT—Fundação para a Ciência e a Tecnologia within project "POCI-01-0145-FEDER-031671".

References

1. Bação, F., Lobo, V., Painho, M.: Applying genetic algorithms to zone design. Soft Comput. **9**, 341–348 (2005)
2. Bouzarth, E.L., Forrester, R., Hutson, K.R., Reddoch, L.: Assigning students to schools to minimize both transportation costs and socioeconomic variation between schools. Socio-Econ. Plann. Sci. **64**, 1–8 (2018)
3. Bozkaya, B., Erkut, E., Laporte, G.: A tabu search heuristic and adaptive memory procedure for political districting. Europ. J. Oper. Res. **144**, 12–26 (2003)
4. Ceyhan, E., Ertuğay, K., Düzgün, Ş.: Exploratory and inferential methods for spatio-temporal analysis of residential fire clustering in urban areas. Fire Safety J. **58**, 226–239 (2013)
5. de Assis, L.S., Franca, P.M., Usberti, F.L.: A redistricting problem applied to meter reading in power distribution networks. Comput. & Oper. Res. **41**, 65–75 (2014)

6. Deb, K., Pratap, A., Agarwal, S., Meyarivan, T.: A fast and elitist multiobjective genetic algorithm: Nsga-ii. IEEE Trans. Evol. Comput. **6**(2), 182–197 (2002)
7. Farughi, H., Tavana, M., Mostafayi, S., Arteaga, F.J.S.: A novel optimization model for designing compact, balanced, and contiguous healthcare districts. J. Oper. Res. Soc. **71**(11), 1740–1759 (2020)
8. Hu, F., Yang, S., Xu, W.: A non-dominated sorting genetic algorithm for the location and districting planning of earthquake shelters. Int. J. Geogr. Inf. Sci. **28**(7), 1482–1501 (2014)
9. Kim, M.J.: Multiobjective spanning tree based optimization model to political redistricting. Spatial Information Research **26**(3), 317–325 (2018)
10. Öztürk, E.G., Rodrigues, A.M., Ferreira, J.S.: Using ahp to deal with sectorization problems. In: Proceedings of the International Conference on Industrial Engineering and Operations Management, pp. 460–471 (2021)
11. Vahdani, B., Veysmoradi, D., Mousavi, S., Amiri, M.: Planning for relief distribution, victim evacuation, redistricting and service sharing under uncertainty. Socio-Econ. Plan. Sci. 101158 (2021)
12. Xu, Z., Liu, D., Yan, L.: Evaluating spatial configuration of fire stations based on real-time traffic. Case Stud. Ther. Eng. **25**, 100957 (2021)

A Holistic Framework for Increasing Agility in a Production Process

Leonor Magalhães, João Sousa Guedes, and Jorge Freire de Sousa

Abstract Following globalization and technological development, markets become progressively more volatile and dynamic, perceiving new competitive advantages related to agility and flexibility. Thus, this study focuses on the development of an agility framework. A project within a leader metal packaging industry worked as the framework trigger. Agility increase was the main target, where production line flexibility leverage and equipment efficiency optimization were also contemplated. However, different barriers to agility were unveiled. Four concepts seem to have an impact on solving the agility problem: Equipment Efficiency, Product Complexity, Portfolio Management, and Production Planning. This paper focuses on how these four forces interact to influence agile manufacturing and understand the methodologies to overcome the problem. The literature presents a gap within this topic that the suggested framework looks to fulfill.

Keywords Agility · Overall equipment efficiency · Production planning · Product complexity · Portfolio variety

1 Introduction

Market volatility demands not only sustainability concerns but also shorter lead times for an offer with a wide variety and level of customization. The agile paradigm is related to market sensitiveness and confers the ability to read and respond to actual demand [8]. The emerging new concepts of agility for companies serving the 21st-century marketplace require the development of a new methodology that focuses not only on machine efficiency but also on broader strategic and disruptive concepts.

L. Magalhães (✉) · J. Freire de Sousa
Faculty of Engineering, University of Porto, Rua Dr. Roberto Frias, s/n 4200-465 Porto, Portugal
e-mail: leonor.gam12@gmail.com

J. Sousa Guedes
Kaizen Institute Western Europe , Rua Manuel Alves Moreira 207, 4405-520 Vila Nova de Gaia, Portugal

J. P. Almeida et al. (eds.), *Operational Research*, Springer Proceedings in Mathematics & Statistics 411,
https://doi.org/10.1007/978-3-031-20788-4_7

Four different areas can separately influence an agility problem, which are Equipment Efficiency, Portfolio Management, Product Complexity, and Production Planning; however, in real world applications, they converge, ascending the need for an integrated end-to-end process. Voice of the Customer (VOC) is the center of the framework since it will define the destination of the road to agility.

Therefore, motivated by the interactions of four areas within a practical experience in the metal packaging industry, the greatest challenge of this paper is developing a holistic model for maximizing agility, enabling a quicker response to market demand without compromising total costs and profit margins.

For this to be possible, it is intended to address the following research questions:

(1) How can each concept impact the agility of a production process?
(2) What are the main interactions between the four concepts?
(3) What is the sequence of the action plan to include in the agility framework?
(4) What are the principal metrics to evaluate the agility target achievement?

The rest of the paper is organized as follows: in the next section, we present the literature review that supports the research carried on. In Sect. 3, we detail the proposed framework on the sequence and methodologies that can improve the agility of a production process. Lastly, we discuss the main conclusions and the future of the research.

2 Literature Review

Driven by different factors, dramatic changes have been occurring in the global market. Many manufacturing companies assist in a tradeoff between efficiency and consumer choice. On one side, customers increasingly want to see their needs met more quickly and cheaply. On the other side, they favor highly customized products with a wide variety of options [15]. Thus, becoming more responsive to the needs of the market is not just about speed; it also requires a high level of agility. In literature, agility is defined as the ability of an organization to respond rapidly to changes in demand, both in terms of volume and variety [6].

Voice of the Customer Voice of the Customer (VOC) is a critical analysis procedure that provides precise information regarding customer input requirements for a product or service output [1]. The main objective of companies is to create products that can represent value for individual customers and attract and convince them to purchase these products. Thus, VOC assumes a fundamental role in driving the product development process [5].

Traditional VOC data collection approaches use questionnaires, telephone surveys, and interviews, which can be time-consuming and costly. Trappey et al. [20] explore processes of collecting VOC from e-commerce platforms, where customers are global. Their comments are reasonably trustworthy since they can only be accessed by customers who have a user experience. Improving the methods of capturing VOC through Research and Development (R& D) leverages the capability to effectively identify changes in customer expectations, which is essential for a company offering short life cycle products in competitive markets. VOC is considered the center of an agility process since customer requirements are the leader factor and must be satisfied for a business to succeed. In a dynamic market, VOC is constantly changing and demanding innovation; thus, it needs to be effectively explored, understood, and examined to overcome uncertainties and fierce market competition.

Portfolio Management Cantamessa [4] defends that product portfolio management is pressured from three main areas: marketing, which involves meeting customer needs; product development, including the complexity of managing multiple product projects; and manufacturing, which is focused on product variety reduction.

To find a good balance between customer satisfaction and portfolio management, it is imperative to understand each distinct customer segment and then develop a concrete offer adapted to individual customer situations and needs [14]. Thus, understanding the VOC should be one of the focuses while building a portfolio.

When multiple product development projects happen simultaneously, the lack of needed resources is one of the fundamental reasons why the products take so long to get to the market. Good product development involves rigorous data—estimates of sales, pricing, manufacturing costs, and development costs—to prioritize which products should enter the portfolio and balance the available resources [9].

Regarding the multiplicity of products within a portfolio, variety can leverage product differentiation; however, it implies higher costs. Production costs get higher since more references make economies of scale difficult. Market analysis costs, too, since managing the supply chain to match fragmented demand is more expensive [4]. The rationalization of Stock Keeping Units (SKUs) is an effective way of reducing product portfolio complexity. However, SKUs influence on firm revenues can limit the impact of this practice [12]. Thus, like Zhu, Shah, and Sarkis [25] defend, product deletion should be well planned to avoid the loss of market segments and revenues associated with the deleted products.

Gregori and Marcone [13], who studied the variety excess within the fashion industry, found that applying a Variety Reduction Program (VRP) with the rationalization of the use of production capacity raises the improved performance of operations, including flexibility of planning and production increase. VRP can also be implemented due to the Bill of Materials (BoM) study, as Cinelli et al. [7] suggest. As a consequence of calculating parts and module numbers within a product portfolio, BoM permits evaluating the similarity between references, accelerating the process of standardization and rationalization of products.

Product Complexity One of the most significant barriers to agility is how complexity tends to increase as companies grow and extend their marketing reach. Often, this complexity comes through product and brand proliferation, being the reduction of product complexity a major priority. Product complexity includes not only design issues (e.g., the number of nonstandard components in a product) but also an excessive variety that does not contribute to more considerable consumer value [6].

According to Orfi, Terpenny, and Sahin-Sariisik [17], manufacturing complex products entails less efficiency, higher setup costs, and the need for more raw material, work-in-process, and finished goods inventory. Thus, the challenges facing the industries today are characterized by design complexity that must be matched with a flexible and complex manufacturing system as well as advanced agile business processes [11]. Varl, Duhovnik, and Tavčar [21] describe the application of agile methods in the re-engineering of a mass customization product. The focus was on reducing complexity and increasing the agility of the design process by introducing modularity in product components. Thus, the same standard design, including common parts and assemblies, process plans, production setups, supply networks, and expertise, could be used in several production scenarios. The result was cost and development time reduction and increased efficiency.

Equipment Efficiency The machine shop floor is the area where a set of sequential processes are carried out to make real what customers want. Efficiency on the machine shop floor is crucial for improved production and effective utilization of all available resources [2]. Overall Equipment Efficiency (OEE) is a quantitative metric that endeavors to identify indirect and 'hidden' productivity and quality costs in the form of production losses [3]. These losses are formulated as a function of availability, performance, and quality.

Availability measures the percentage of time the equipment or operation was running compared to the available time. Performance evaluates the operation speed compared to its maximum capability, and the quality term refers to the number of good parts produced compared to the total number made [16]. Besides being used as a performance measure, OEE is also used as an indicator of manufacturing issues, such as excessive breakdowns, lack of preventive actions, and an effective corrective approach [10]. Finding the right ways to improve equipment efficiency is one of the tools to achieve greater manufacturing flexibility for smaller series traduced in quicker market response.

Production Planning Production planning can interact with agility through three main metrics: lead time, Every Part Every Interval (EPEI), and Minimum Order Quantity (MOQ).

Mukherjee, Sarkar, and Bhattacharjya [19] reveal that, in MTO systems, a product is prepared from the component as per the specification of the customer's order; in MTS, lead time is shorter as the customer's order is fulfilled from stock. The authors also notice that finished product inventory in MTS increases with product variety, based on the anticipation of customer demand. Every production system should be MTO with a pull strategy in an ideal status. Production is pulled from the customer's request and the lead time is so short that inventory is not necessary, building a zero-waste value stream. This process should be lean due to the total percentage of value within the process and agile since the production keeps up with a volatile market. Production planning affects agility through lead time. In MTO mode, although production activity could only start after the orders were received, customers do not want to wait for a long time; hence, a short lead time is a must. The concept of EPEI describes the overall time in which all product variants can be produced on one defined resource [18]. EPEI metric is especially important for production processes with a wide variety of products in a repetitive production environment, reflecting the flexibility of the value chain. A low value of EPEI means there is a high rotation of production within a manufacturing line, and consequently, an increased agility level is associated.EPEI is the ratio between the number of product references to produce and the number of setups available within the production line in a determined interval. If the EPEI is higher than delivering the lead time promised to the customer, inventory of product references will be necessary to satisfy customer demand, which is considered waste within lean production. Thus, the MOQ required per production order will be higher since there is a need to stock.

Interactions Analysis In summary, the agility of a process is related to responding with high velocity to fluctuations in demand, which tend to be unstable and demand great variety. Thus, industries' manufacturing processes must be efficient to keep up with the increasing market requirements they face nowadays. A product portfolio full of variety implies faster setups and a high production line efficiency to allow a more significant rotation within product references and guarantee a low volume of stock, which is the key to lean production.

With the alignment between product portfolio and equipment efficiency, flows a well-planned production, characterized by satisfying the demanded short lead times and producing the minimum inventory levels. When possible, complexity in the value chain must be eliminated to achieve this flexible status. Formerly, product complexity was addressed as a barrier to agility because it hinders all related processes, with alternatives like modularity.

Based on the literature review, conclusions about the relationships between the covered topics can be taken. The four different areas are all connected. However, for example, as the reduction of portfolio complexity positively impacts equipment efficiency, the reverse does not occur. In the case of several references abundance, the higher manufacturing efficiency will not affect the portfolio optimization, leading to a slow achievement of the agile purpose. In this sense, Table 1 exposes all the one-to-one dependencies of each relationship and how they impact each other.

A model capable of including all these interactions has not been presented yet and could be useful in managerial practice. However, the literature already includes methodologies for achieving agility in manufacturing organizations. Zhang and Sharifi [24] developed a conceptual approach for implementing agility in the industrial context based on tools to help strategic decision-making within the agile manufacturing scope. The first step arises with the need to define the environmental pressures that influence the competitive advantage of a business. Then, it is necessary to find the companies' mandatory capabilities for positively responding to those pressures. After realizing the critical influence factors and the aspired capabilities, the last step concerns the means for achieving them.

Vásquez-Bustelo, Avella, and Fernández [22] consider the presence of a turbulent business environment as a trigger to develop an agile manufacturing process. The dynamic environment is essentially related to four areas: the unpredictable changes in the environment, the high market competition, the close links between firms and all the stakeholders in the supply chain, and the high diversity of products, lines, and customers within businesses. Vázquez-Bustelo, Avella, and Fernández [22] then

Table 1 Cause-effect relationships review

$$PorfolioManagement \underset{Noimpact}{\overset{Impact}{\rightleftarrows}} EquipmentEfficiency$$

$$PorfolioManagement \underset{Noimpact}{\overset{Impact}{\rightleftarrows}} ProductionPlanning$$

$$PorfolioManagement \underset{Impact}{\overset{Noimpact}{\rightleftarrows}} ProductComplexity$$

$$ProductComplexity \underset{Noimpact}{\overset{Impact}{\rightleftarrows}} EquipmentEfficiency$$

$$ProductComplexity \underset{Noimpact}{\overset{Impact}{\rightleftarrows}} ProductionPlanning$$

$$EquipmentEfficiency \underset{Impact}{\overset{Impact}{\rightleftarrows}} ProductionPlanning$$

created a model and tested it in a sample of the largest manufacturers in Spain with a structural equation model application to check determined hypotheses. The multidimensional nature of agile manufacturing was confirmed by drawing up a measurement scale based on the integration of both structural and infrastructural manufacturing practices.

The present study identifies the four listed factors as the major pressures that can influence a production process and be critical in the pursuit of agile manufacturing. These pressures were considered urgent when presenting a lack of optimization for achieving an agile process.

3 Agility Framework

Based on the relationships of impact between each of the four concepts, both the sequence and the methodologies for leveraging the agility of a production process were built, being systematized in Table 2.

The approach is firstly based on complexity analysis, in which product complexity and portfolio management arise. VOC is an input of the whole framework since it is fundamental to understand customers' requirements concerning the offered product.

Product complexity evaluation is suggested in an earlier stage than the portfolio itself since reducing difficulties within product scope can influence reference elimination and highlight some hidden portfolio problems. Only after cutting the non-value-added complexity from the value stream, does the process become ready for an equipment efficiency improvement. Production planning is represented as embracing all the model sequences since the impact within lead time, the minimum production quantities, and the leverage of the production sequence is progressive during framework implementation.

Table 2 Agility framework: proposed sequence and methodologies

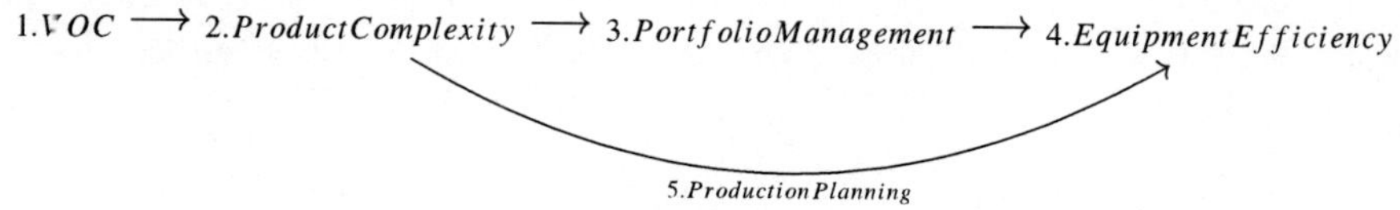

	Methodologies	Metrics
Product complexity	VAVE	Number of components
Portfolio management	VRP	Number of references
Equipment efficiency	SMED, Kobetsu	OEE, Setup time
Production planning	Pull planning	EPEI, Lead time, MOQ

Since the present study is based on one of Kaizen's consulting projects, the methodologies proposed by the Agility Framework are the most used by the company and, therefore, with a high success rate over time. However, diverse methods and metrics can also be used, being the sequence of the different focus areas the most significant constraint for achieving the results.

Product Focus Different methodologies can be used to analyze product scope and create actions to reduce its complexity. The proposed framework is centered on VAVE.

Value Analysis Value Engineer (VAVE)
VAVE focuses on value improvement through an ongoing critical examination of current product features, reducing product cost, and maximizing product value. Thus, deep VOC knowledge is crucial for its success before applying this method, since the decisions will be based on what adds true value to the consumer.

Understanding VOC in an iterative way, i.e., with constantly updated market information related to customers' preferences and expectations, will help identify critical decision-making characteristics. VOC can be collected in different ways, like surveys and market research, one-on-one questionnaires, customer feedback through social media, or other marketing activities. Therefore, it is critical to understand the VOC requirement when agility is the focus.

VAVE workshop is divided into Value Analysis (VA) and Value Engineer (VE) scopes. VA is based on evaluating the existing product specifications and finding opportunities to reduce complexity without compromising value delivery. The VE topic stands out when it is possible to reduce costs within the current features. VE is about disruptive innovations, and process creativity to leverage product development and lead it to a higher value level.

Modularity is a topic that is gaining expression through literature and can perfectly fit into the VAVE methodology. Modularity, described as involving cost reduction by including coupling features and reducing the number of parts, leverages the value of the process without interfering with the customer experience. Standardization of

Table 3 Variety reduction program steps adapted from Kaizen Institute, [23]

Value stream design	Cost model	Reference elimination	Team organisation
Identify where the complexity is created	TDABC- Time-driven activity based costing	Define the list of SKUs to keep or eliminate	Aligned production and commercial departments
Step 1	Step 2	Step 3	Step 4

parts between products, for example, through common parts or sub-assemblies, can reduce product complexity and costs.

VAVE should involve a multidisciplinary team to be effective, where procurement engineers should assume an active role, including product estimating tasks, and where development and purchasing members should also participate. Concrete ideas should arise as an output of the method, with a first estimate of the associated potential benefits. These must be driven by clear actions and responsible teams or individuals for each task, as well as planned timings. To succeed, generated ideas must be put into practice as soon as possible to confirm the gains.

Portfolio Focus After attacking the constraints within the product dimension, it is necessary to consider the group of the product offered next. Product portfolio rationalization can significantly impact agility increase, being one of the causes of lower values of process efficiency and difficulties within production planning. However, references should be eliminated through a careful process, which must consider the tradeoff between revenues each product brings versus its total complexity costs. One of the tools used in this sense is the Variety Reduction Program (VRP), developed to reduce complexity without compromising total margins and customer offer satisfaction and value delivered.

Variety Reduction Program (VRP)
VRP is a methodology with the scope of optimizing the number of SKUs to increase a firm's margin and sales, leverage its service level, and promote a better alignment between production and commercial portfolio. Greater simplicity and flexibility increase the odds of higher performance; thus, VRP helps leverage the harmony within the portfolio through the four steps described in Table 3.

VRP output should be a sustainable portfolio in which both production and commercial departments connect, and product development is done wisely. Excluding references that skewed production processes and did not add superior value to the customer opens doors for agile manufacturing. However, manufacturing activity must be prepared and optimized in that way, corresponding to the next step of the framework.

The total number of references within a portfolio can work as a metric for the VRP process. However, it is also essential to constantly evaluate if process capacity adapts to the current list of references and if there are enough resources to achieve greater agility.

Efficiency Focus Problems related to machine efficiency in a production process can connect to any OEE component: availability, performance, and quality. An agile

Table 4 Five steps of SMED adapted from Kaizen, [23]

Work study	Separate internal from external work	Convert internal work into external work	Reduce internal work	Reduce external work
Step 1	Step 2	Step 3	Step 4	Step 5

process premises an efficient machine or line changeover when manufacturing a different reference is required. In that sense, availability losses through setups can be solved by applying the SMED methodology. Regarding specific machine problems, due to higher breakdown occurrences or quality issues, the Kobetsu method is the most indicated.

Single Minute Exchange of Die (SMED)
SMED is a methodology focused on reducing unproductive change over time, making setups flexible to enable manufacturing agility. After VRP, products left in the portfolio are the most valuable ones to the business, being essential to grant the minimum time loss within setups.

As shown in Table 4, the first step of the method is to register the setup through video and define each operator's tasks and their respective duration. After analyzing the As-Is status, it is crucial to distinguish work that needs to be done with the machine stopped, internal work, from work that can be done with the machine working, external work. Thus, every task, like cleaning and finding or grabbing necessary tools, is reorganized before or after machine stoppage. In the third step, the main gains correspond to converting what is currently internal to external work, such as introducing pre-assemblies, pre-adjustments, or pre-heating processes, which can save setup time.

Reducing internal work consists of finding solutions to facilitate operators' tasks during the setup, like simplifying fixations and tightening, introducing parallel work by reorganizing participant tasks, or even creating standards. Finally, after applying the previous steps, quick wins can be already notable and maybe even more evident if the external work is optimized. The fifth step consists of improving support logistics and serviceability so that tasks like the operator's movements, which do not bring value, could be eliminated. SMED output can be measured through OEE and setup duration, which will leverage process availability.

Kobetsu Kaizen
Kobetsu is a machine-focused method typically applied to the critical machines with a lower OEE. Kobetsu's approach begins with mapping the current process and, after defining the desired status, studying the root causes for the efficiency issue. Having a clear map of the process, solutions that emerge through brainstorming with the participants need to be implemented. Results are then tested, and solutions can iterate if necessary.

An important tool used in a Kobetsu event is the Failure Modes and Effects Analysis (FMEA), to be included in the initial evaluation of the As-Is situation.

When the detected problems are related to high breakdown losses, evaluating the machine components, their probability of failure, and respective impact is valuable to define concrete actions to implement. From an FMEA, for every failure mode and corresponding effect, its severity, frequency, and the propensity of detection are determined, resulting in a risk priority classification. The failures with the highest risk levels must have superior attention, and recommended actions for those problems must be conducted.

Kobetsu action plan can reach preventive and autonomous maintenance topics since the lack of these regular measures can be the root cause for poor machine efficiency. Maintenance should be planned in the most critical machine components to increase equipment reliability and its lifetime. Moreover, practices of autonomous maintenance, which are the responsibility of the operators, should be included in their regular tasks. Not only basic cleaning and machine lubrification are included in autonomous maintenance, but also utilization of checklists and verification plans, which will leverage machine efficiency while reducing their probability of breakdown failure.

Equipment efficiency focus will then result in the OEE increase, which reduces setup time and leverages OEE's components. After implementing the previous framework methodologies, it is expected that the achievement of an agility increase through the process will be reflected in the production planning scope.

Production Planning Focus Within the agility framework, production planning is continuously represented along the three other stages due to its consequent improvement facing the implemented methodologies. Production planning is related to lead time, MOQ, and EPEI metrics, which will constantly be improved throughout the road to agility.

After the first stage, the production process is free of product complexity, resulting in an optimized product cost that delivers the whole required value to the customer. By reducing material costs, introducing modularity concepts, and reducing the number of product components, processes get more straightforward, which can already impact lead time. Following portfolio management focus, as the less valuable and more complex references are eliminated, manufacturing becomes more agile due to the reduction of SKUs and the number of needed changeovers. Thus, having fewer setups required, the production process gains efficiency and reduces EPEI, leveraging production rotation. Therefore, by the end of the second stage, manufacturing experiences a flexibility improvement due to the shorter interval in which the same products are produced, allowing a MOQ reduction.

As explored before, EPEI can be improved not only by reducing the number of portfolio references but also by increasing OEE and reducing setup time. By being centered on machines with a critical impact on OEE and working on solutions for the actual root causes, Kobetsu can contribute to optimizing equipment efficiency. Moreover, the SMED methodology, which has the primary goal of reducing setup time between production references, allows a significant reduction of stoppage time during changeovers. By the end of the third framework stage, EPEI is minimized. The production process achieves a flexible and agile status, where the manufacturing

Table 5 Pull planning implementation sequence adapted from Kaizen, [23]

Strategic planning	Capacity planning	Execution planning
–Make to order	–Equipment efficiency	–Customers' orders
–Make to stock	–Logistics	–Replenishment orders
Step 1	Step 2	Step3

interval between the portfolio references is the shortest. Thus, it is now viable to produce smaller orders due to the flexibility of the process, which is associated with a shorter lead time.

During framework implementation, production management gets less complicated in what concerns the production sequence since SKU selection is optimized and efficiency constraints are overpassed. Even if production planning improvement is a consequence of the previous methodologies, a pull planning strategy is a way to reach a superior state of agility.

Pull Planning

Pull planning strategy is based on creating flow within logistics and production, having a lean system where the material replenishment is pulled by customer demand, so less stock is needed. This strategy can be applied both in MTS and MTO systems. MTO is associated with higher lead times than MTS due to the nonexistence of stock, but the idea of agility is better traduced in an MTO system. MTO value stream combined with the agility road is characterized by low inventory and shorter lead times due to low EPEI levels and high manufacturing efficiency. Thus, even a product with high demand variability, which typically implies the existence of stock, can be provided by an MTO system, pulled by customer's orders, if the agility of the process is optimized.

The idea of pull planning is a management strategy based on actual demand and not on forecasts. As represented in Table 5, the first step to implementing pull planning is defining product strategy: the strategy of each reference depends on its demand and the cost of holding the final product in inventory. Products with a low level of sales and a low cost of stock may be worth assuming an MTS strategy.

Then, the supply chain must be dimensioned according to production capacity to satisfy customers' needs.Logistics must guarantee the replenishment of materials required at the exact time and place and in the right quantity. On the other hand, production must ensure enough capacity and space for intermediate stock or materials. An alignment between production and logistics is a requirement for flow efficiency. Execution planning arises as a consequence of the defined strategy and capacity planning. For MTO products, the execution plan is based on customer orders, yet for MTS products, the plan is based on replenishment orders. Stock volume per SKU must be constantly updated based on lead time and EPEI since the inventory corresponds to the average quantity needed in the interval between producing the same reference plus the production lead time; this leads to a service level of excellence.

A flexible process can now harmoniously plan its production with a volatile demand as an input and the challenge of responding quickly to smaller quantities and shorter lead times.

4 Conclusions and Future Research Directions

A generic framework was built to describe the road to agility when process difficulties belong to the interaction of Portfolio Management, Product Complexity, Equipment Efficiency, and Production Planning. Table 6 represents the generic issues that characterize an agility problem and fit in the agility framework. It both systematizes the specific dimension issue, as well as the generic characteristics that concern the relation between the two dimensions.

A holistic sequence is suggested in which the first step is studying the VOC to eliminate complexity, both on product and portfolio dimensions. The idea is to deliver the same value to the customer. Then, after removing any product characteristic with no value and evaluating which references to cut from the product portfolio, a large number of product components with no value-added and an abundant number of products stop being an issue.

Even though the portfolio simplicity requires a low number of production setups, those line interruptions need to be optimized in order to achieve a higher OEE and lower stock values. Consequently, the next focus is on equipment efficiency.

Following the implementation sequence as suggested in the previous section, improvements at the production level will arise, and the interactions between each

Table 6 Generic matrix of characteristics that fit into the agility problem

	Portfolio management	Product complexity	Equipment efficiency	Production planning
Portfolio management	Great number of references			
Product complexity	Redundancy of products	Great number of components with no value–added		
Equipment efficiency	Great number of setups	Product features that bring complexity to production process	–High Setup times –Low OEE	
Production planning	–High levels of stock –Complexity in production sequence	Early point of differentiation in the production process	High EPEI	–High lead times –High MOQ

factor will be solved. As complexity is eliminated and OEE increases, with a perfect adequation of available resources to customer demand, the lead time will decrease, as well as minimum order quantities allowed. After the framework implementation, the process will be more agile and flexible, ready to respond quickly to market fluctuations.

The developed framework should be implemented from scratch in an industrial context in future research. The success of the prospective case study should be measured both in terms of equipment efficiency and of all the active topics. Product complexity reduction must be evaluated through the number of product components and product costs. The number of product references must track the optimization of the product portfolio. OEE and setup time evolution must measure the equipment efficiency. Production planning must be evaluated through EPEI, lead time, and MOQ. Production Planning KPIs are critical to analyzing the agility achievement status since the target is traduced in short lead times, low MOQ, and low EPEI. Thus, the future project will focus on the whole value stream.

Generically, all the research questions outlined in the beginning were responded to through the article. It is clear how each addressed concept can impact the agility of a value stream and how they can interact with each other. The principal issues of a generic agility problem were traduced as well as the suggested action sequence to overcome the flagged difficulties. The developed road to agility can support companies in understanding the dynamic of agility and how to achieve it by applying the proposed methodologies.

References

1. Aguwa, C.C., Monplaisir, L., Turgut, O.: Voice of the customer: customer satisfaction ratio based analysis. Expert Syst. Appl. **39**(11), 10112–19 (2012). https://doi.org/10.1016/j.eswa.2012.02.071
2. Ajay Guru Dev, C., Senthil Kumar, V.S., Rajesh, G.: Effective human utilization in an original equipment manufacturing (OEM) industry by the implementation of agile manufacturing: a POLCA approach. Hum. Factors Ergon. Manuf. Serv. Ind. **27**(2), 79–86 (2017). https://doi.org/10.1002/hfm.20692
3. Aminuddin, B., Ainunnazli, N., Garza-Reyes, J.A., Kumar, V., Antony, J., Rocha-Lona, L.: An analysis of managerial factors affecting the implementation and use of overall equipment effectiveness. Int. J. Prod. Res. **54**(15), 4430–47 (2016). https://doi.org/10.1080/00207543.2015.1055849
4. Cantamessa, M.: Product portfolio management. In: Design Process Improvement: A Review of Current Practice, 404-35. Springer London (2005). https://doi.org/10.1007/978-1-84628-061-0_18
5. Carulli, M., Bordegoni, M., Cugini, U.: An approach for capturing the voice of the customer-based on virtual prototyping. J. Intell. Manuf. **24**(5), 887–903 (2013). https://doi.org/10.1007/s10845-012-0662-5
6. Christopher, M.: The agile supply chain: competing in volatile markets. Ind. Mark. Manage. **29**(1), 37–44 (2000). https://doi.org/10.1016/S0019-8501(99)00110-8
7. Cinelli, M., Ferraro, G., Iovanella, A., Lucci, G., Schiraldi, M.M.: A network perspective for the analysis of bill of material. In: Procedia CIRP, 88:19–24. Elsevier B.V. (2020). https://doi.org/10.1016/j.procir.2020.05.004

8. Collin, J., Lorenzin, D.: Plan for supply chain agility at Nokia: lessons from the mobile infrastructure industry. Int. J. Phys. Distrib. Logist. Manag. **36**(6), 418–30 (2006). https://doi.org/10.1108/09600030610677375
9. Cooper, R.G., Sommer, A.F.: New-product portfolio management with agile: challenges and solutions for manufacturers using agile development methods. Res. Technol. Manag. **63**(1), 29–38 (2020). https://doi.org/10.1080/08956308.2020.1686291
10. Dal, B., Tugwell, P., Greatbanks, R.: Overall equipment effectiveness as a measure of operational improvement—a practical analysis. Int. J. Oper. Prod. Manag. **20**(12), 1488–1502 (2000). https://doi.org/10.1108/01443570010355750
11. Elmaraghy, W., Elmaraghy, H., Tomiyama, T., Monostori, L.: Complexity in engineering design and manufacturing. CIRP Ann. Manuf. Technol. **61**(2), 793–814 (2012). https://doi.org/10.1016/j.cirp.2012.05.001
12. Campos, F., Pablo, P.T., Huatuco, L.H.: Managing structural and dynamic complexity in supply chains: insights from four case studies. Prod. Plan. Control **30**(8), 611–23 (2019). https://doi.org/10.1080/09537287.2018.1545952
13. Gregori, G.L., Marcone, M.R.: R&D and manufacturing activities regarding managerial effectiveness and open strategy: an industry focus on luxury knitwear firms. Int. J. Prod. Res. **57**(18), 5787–5800 (2019). https://doi.org/10.1080/00207543.2018.1550271
14. Heikkilä, J.: From supply to demand chain management: efficiency and customer satisfaction. J. Oper. Manag. **20**(6), 747–67 (2002). https://doi.org/10.1016/S0272-6963(02)00038-4
15. Ismail, H., Reid, I., Poolton, J., Arokiam, I.: Mass customization: balancing customer desires with operational reality. Int. Ser. Oper. Res. Manag. Sci. **87**, 85–109 (2006). https://doi.org/10.1007/0-387-32224-8_5
16. Nayak, D.M., Sreenivasulu Naidu, G., Shankar, V., Manager, A., Manager, A.: Evaluation of OEE in a continuous process industry on an insulation line in a cable manufacturing unit. Int. J. Innov. Res. Sci., Eng. Technol. **2**(5), 1629–34 (2013)
17. Orfi, N., Terpenny, J., Sahin-Sariisik, A.: Harnessing product complexity: step 1 establishing product complexity dimensions and indicators. Eng. Econ. **56**(1), 59–79 (2011). https://doi.org/10.1080/0013791X.2010.549935
18. Sihn, W., Pfeffer, M.: A method for a comprehensive value stream evaluation. CIRP Ann. Manuf. Technol. **62**(1), 427–30 (2013). https://doi.org/10.1016/j.cirp.2013.03.042
19. Mukherjee, K., Sarkar, B., Bhattacharjya, A.: Supplier selection strategy for mass customization. In: 2009 International Conference on Computers and Industrial Engineering, CIE 2009, 892-95. IEEE Computer Society (2009). https://doi.org/10.1109/iccie.2009.5223861
20. Trappey, A.J.C., Trappey, C.V., Fan, C.Y., Lee, I.J.Y.: Consumer driven product technology function deployment using social media and patent mining. Adv. Eng. Inform. **36**(April), 120–29 (2018). https://doi.org/10.1016/j.aei.2018.03.004
21. Varl, M., Duhovnik, J., Tavčar, J.: Agile product development process transformation to support advanced one-of-a-kind manufacturing (2020). https://doi.org/10.1080/0951192X.2020.1775301
22. Vázquez Bustelo, D., Avella, L., Fernández, E.: Agility drivers, enablers and outcomes: empirical test of an integrated agile manufacturing model. Int. J. Oper. Prod. Manag. **27**(12), 1303–1332 (2007)
23. Imai, M.: Gemba Kaizen, a commonsens approach to a continuous improvement strategy. Kaizen Institute, Ltd. (2020)
24. Zhang, Z., Sharifi, H.: A methodology for achieving agility in manufacturing organisations. Int. J. Oper. Prod. Manag. **20**(4), 496–513 (2000). https://doi.org/10.1108/01443570010314818
25. Zhu, Q., Shah, P., Sarkis, J.: Addition by subtraction: integrating product deletion with lean and sustainable supply chain management. Int. J. Prod. Econ. **205**, 201–214 (2018). https://doi.org/10.1016/j.ijpe.2018.08.035

Nesting and Scheduling for Additive Manufacturing: An Approach Considering Order Due Dates

Paulo Nascimento, Cristóvão Silva, Stefanie Mueller, and Samuel Moniz

Abstract This paper proposes a new Constraint Programming (CP) formulation to solve the problem of nesting and scheduling parallel additive manufacturing (AM) machines while minimizing operating and tardy deliveries costs. To the best of our knowledge, this is the first model of its type that integrates nesting and scheduling decisions while considering irregular-shaped parts. Considering tardy deliveries, and consequently due dates, allows balancing the fulfillment of order due dates with machines' capacity utilization, opposed to similar studies that tend only to consider the minimization of makespan, not guaranteeing the fulfillment of due dates. The CP model can be solved efficiently, allowing the study of different degrees of flexibility in terms of job allocation and sequencing. Different instances are used to demonstrate the applicability and performance of the proposed formulation.

Keywords Additive manufacturing · Scheduling · Nesting · Bin-packing · Constraint programming

1 Introduction

Additive manufacturing (AM) processes are suitable to deliver high-variety, low-volume products on-demand, and AM factories are typically composed of multiple parallel printers [1]. Hence, to achieve high production efficiency, adequate AM scheduling methodologies are crucial [2]. This paper addresses this problem by proposing a new constraint programming (CP) approach. Opposed to the few studies in this field that intend to minimize makespan, our goal is to minimize the total

P. Nascimento (✉) · C. Silva · S. Moniz (✉)
Mechanical Engineering Department, University of Coimbra, Coimbra, Portugal
e-mail: pnascimento@student.dem.uc.pt

S. Moniz
e-mail: samuel.moniz@dem.uc.pt

S. Mueller
MIT CSAIL, Cambridge, USA

J. P. Almeida et al. (eds.), *Operational Research*, Springer Proceedings in Mathematics & Statistics 411,
https://doi.org/10.1007/978-3-031-20788-4_8

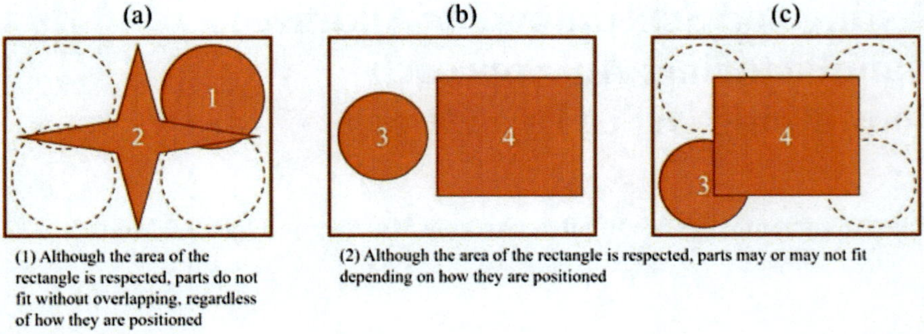

Fig. 1 Examples demonstrating the need for a packing procedure

costs resulting from machines' utilization and tardy deliveries. To the best of the authors' knowledge, this is the first exact approach that adequately addresses the AM scheduling problem by considering the packing of irregular-shaped parts in the machines' building platforms.

On the problem structure, considering that multiple parts (different or not) may be produced in one run, the scheduling of AM machines is a variant of the Batch Processing Machine (BPM) scheduling problem [3]. However, most of the works addressing this setting consider the machine's capacity as an one-dimensional knapsack constraints [4]. This means that the capacity is simplified to a value corresponding to the maximum weight, volume, or area the machine can support in one batch. AM, in turn, brings another degree of complexity because the machine's capacity is represented at least by a two-dimensional rectangle, where each part will occupy a specific zone of its inner area without overlapping with other parts.

Three different examples are presented in Fig. 1, where the outer rectangles represent the AM's machine building platform, and the inner objects represent the parts to be produced. In Fig. 1a, although the total area of the two parts respects the area of the outer rectangle, the shape of part 2 makes it unfeasible to place both parts without overlapping. In Fig. 1b and Fig. 1c, the two parts whose total area respects the area of the outer rectangles may or may not fit depending on the packing configuration. In this sense, only considering the area for constraining the machine capacity, disregarding how parts are positioned and their shape, may lead to infeasible solutions.

The capacity in AM needs then to be modeled as a nesting problem (2D bin-packing problem). The problem consists of packing multiple parts in a certain number of rectangles that correspond to batches constrained by the capacity of the AM machines building platforms. In this sense, the AM scheduling problem tackled in this paper combines the BPM scheduling problem with the nesting problem. Both problems are proven to be NP-hard, and thus, it is possible to conclude that the AM scheduling problem is strongly NP-hard [5]. Our work proposes to tackle nesting and scheduling decisions from an integrated perspective while considering irregular-shaped parts. Solving these two problems simultaneously is very important since nesting decisions can change the solution space of the scheduling problem and vice versa.

For simplicity, we will adopt the terminology suggested by Oh et al. [2] in the rest of this work. Part refers to an object that needs to be produced (similar to job in the scheduling literature). Build consists of a group of parts that are simultaneously produced in the same building platform (similar to batch in the scheduling literature). Nesting consists of grouping parts into builds while determining their location within the build (similar to 2D bin-packing).

The rest of the paper is structured as follows. Section 2 presents a brief literature review on AM technologies and approaches for BPM scheduling, nesting, and AM scheduling problems. Section 3 describes the problem statement. Section 4 presents in detail the developed model. Then, the results obtained by the model are analyzed in Sect. 5. Lastly, conclusions and future studies are presented in Sect. 6.

2 Literature Review

This section discusses the AM technological context relevant to the present problem and critically reviews the most relevant model-based approaches.

2.1 AM Characteristics Relevant for Scheduling and Nesting Decisions

A possible setting of the AM machines is the ability to stack parts up. In these cases, the machine's capacity must be represented as a three-dimensional cuboid, resulting in a 3D bin-packing problem. In this work, we focus on an integrated analytic model considering 2D bin-packing decisions as, in practice, many printers work in a two-dimensional plane without stacking.

Another essential characteristic is related to the processing time of each build. In AM, the processing time of a build is the sum of the processing times of each layer. However, while for some technologies the processing time of a layer can be independent of the number and shape of parts, in other technologies this dependency exists. Although our model can be extended to the second case, we focus on the first case.

2.2 Nesting and Scheduling Integration

AM technology is suitable for customized production environments, therefore the structure of the scheduling problem can be defined as a BPM scheduling problem with non-identical jobs. This problem was first introduced by Uzsoy [6]. Later on, Arroyo & Leung [7] developed a mixed-integer programming (MIP) model to the same problem but assumed non-identical parallel machines, while Ham et al. [8] studied the identical parallel machines version and demonstrated that a CP approach can be very efficient.

Regarding the nesting problem, and contrasting to the high number of available heuristics and metaheuristics, few exact methods exist to solve this problem. Toledo et al. [9] proposed a MIP approach where the placement area is represented by a mesh of dots. Considering concepts of no-fit polygon (NFP) and inner-fit polygon (IFP), parts are placed in those dots without overlapping. Cherri et al. [10] extended this work by proposing three new approaches based on CP where parts can be placed in multiple placement areas. The work of Cherri et al. [10] seems then to be the most advanced work within exact approaches for nesting irregular-shaped parts. Our model is partially based on the ideas of their work.

Next, we provide a relatively small body of literature specific to AM scheduling. Kucukkoc et al. [11] was the first work to address AM scheduling for parallel machines, developing a MIP model, and several heuristic procedures. Later on, the same authors presented an improved model by making it more efficient [12]. The main limitation of these works is that the process of grouping parts together into builds is done considering the total area of the AM building platform while ignoring placement issues. More recently, Che et al. [3] proposed a mixed-integer linear programming (MILP) model and a simulated annealing algorithm to tackle this problem. In this work, although nesting is considered, parts are reduced to their rectangular bounding box. Zhang et al. [4] presented an evolutionary algorithm for AM scheduling, and this is the only work found that adequately addresses the nesting of irregular-shaped parts without significant simplifications. Something common to all previously mentioned studies is that due dates are ignored.

Since no exact approaches were found that adequately addressed AM scheduling without relevant simplifications in the process of grouping parts into builds and due dates are typically ignored, we focus on developing a model-based approach to tackle these gaps.

3 Problem Description

We consider multiple parts nesting and scheduling in AM parallel machines, taking into account parts' delivery times to fulfill customers' due dates. Those parts, indexed by $i = 1, 2, \ldots, n_i$, will be scheduled in multiple identical parallel AM machines, indexed by $m = 1, 2, \ldots, n_m$. Parts, whose individual demand is considered to be one, are grouped into builds, indexed by $b = 1, 2, \ldots, n_b$, and allocated to one of the available machines. Each build will then be produced according to the schedule of the respective AM machine. Thus, our problem is to group the different parts into builds and allocate them to one of the available machines while respecting the capacity of the building platform.

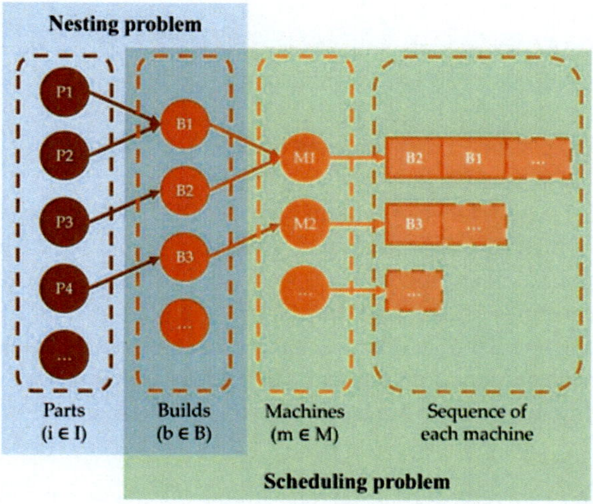

Fig. 2 Illustration of the AM scheduling problem

According to the taxonomy presented by Oh et al. [2], the problem tackled here belongs to the class [M/M/iM]—Multi-part, multi-build, identical multi-machine. Figure 2 provides a visual representation of our problem.

Important assumptions are that the processing time of each build is assumed to be dependent on the tallest part produced, i.e., the part with the longest processing time, all parts are considered to fit in the building platform of the AM machines, and this study focuses on a 2D placement, which means that parts are not allowed to stack up.

4 The Constraint Programming Model

CP is an attractive approach as it can obtain good quality solutions fast. CP is very well known in the field of artificial intelligence and is a prominent modeling technique with proven success in efficiently solving nesting [10] and scheduling [8] problems. We developed our model in the CP optimizer provided by IBM ILOG CPLEX Optimization Studio that offers built-in functions and constraints for modeling scheduling problems. The model notation presented next follows the CPLEX syntax.

4.1 Notation

The rest of this subsection describes and explains the sets, subsets, parameters, and variables that compose the CP model.

Sets

$i \in I$—set of parts, indexed by $i = 1, \ldots, n_i$
$m \in M$—set of AM machines, indexed by $m = 1, \ldots, n_m$
$b \in B$—set of builds, indexed by $b = 1, \ldots, n_b$
$d \in D$—set of dots, indexed by $d = 1, \ldots, n_d$

Each build b can contain various parts i and will be assigned to a specific machine m. A build is characterized by a mesh of dots, and each part has a positioning point which is then placed in a specific dot of the build. Considering that the builds are rectangles of the same dimensions, the total number of dots n_d is dependent on the total number of builds n_b and the total number of dots per build. The value of n_b refers to the number of builds and will be discussed in Sect. 5.2. For a better understanding of how nesting is done see the work of Cherri et al. [10].

Subsets

IFP_i—set of dots of the inner-fit polygon between part i and the builds
$NFP_{i\,i'}^{d}$—set of dots of the no-fit polygon between part i and i' if part i is placed on dot d

These subsets are crucial as the model uses them to determine where and how parts can be positioned, and this is how the model can deal with different parts regardless of their shape. Since this is a time-consuming process, these subsets are calculated in advance. For details about the subsets IFP_i and $NFP_{i\,i'}^{d}$ see the work of Toledo et al. [9].

Parameters

h_d—build to which dot d belongs
p_i—processing time of part i
r_i—release time of part i
d_i—due date of part i
NB—fixed cost of each build used
PC—processing cost of an AM machine per unit time
TD—cost of delivering a part after its due date
MD—maximum delay of a part

Binary Variables

X_{id}—equals to 1 if the positioning point of part i is assigned to dot d, 0 otherwise
T_i—equals to 1 if part i is delivered after its due date, 0 otherwise

Interval Variables

Interval variables are intrinsic variables of the CP optimizer dedicated to scheduling, and represent an interval of time during which an activity occurs. These variables are defined by a start time, a processing time, and an end time, where the start time plus the processing time equals the end time. Thus, in the case of Y_b, it provides the start, processing, and end time of build b. A key feature is that interval variables can be optional, which is why Y_b only exists if build b is used, and N_{bm} only exists if build b is used and assigned to machine m.

Y_b—optional interval variable that represents the interval of time during which build b is processed (Y_b optional in $[\min_{i \in I} r_i, \max_{i \in I} d_i + \text{MD}]$)
N_{bm}—optional interval variable that represents the interval of time during which build b is processed in machine m (N_{bm} optional size $[\min_{i \in I} p_i, \max_{i \in I} p_i]$)

Sequence Variables

Sequence variables are also inherent variables of the CP optimizer dedicated to scheduling, and their function is to determine the order of a set of interval variables, which in this case are the variables N_{bm} of each machine m. Intrinsically, variables N_{bm} that do not exist, either because build b is not used or is not assigned to machine m, will not be considered in the ordering.

S_m—order of the builds (intervals of time) assigned to machine m (S_m *in* $all(b\ in\ B)\ N_{bm}$)

4.2 Model Formulation

4.2.1 Objective Function

$$min\left(NB \times \sum_{b \in B} presenceOf(Y_b) + PC \times \sum_{b \in B} sizeOf(Y_b) + TD \times \sum_{i \in I} T_i\right) \quad (1)$$

The objective is to minimize a total cost function consisting of fixed and variable operating costs and the cost associated with late deliveries. In short, the first term accounts for the fixed build costs. This is done by checking if build b is used through the built-in function of the CP optimizer $presenceOf(Y_b)$, which returns 1 or 0 depending if build b is used or not, respectively. The second term computes the variable utilization costs, and the third term enforces the minimization of the costs related to the tardy deliveries. Considering a cost penalty for late deliveries is interesting since AM processes aim to run in a just-in-time fashion. Therefore, we can assume that measuring the number of tardy jobs is relatively more important than accounting for the jobs lateness. Nonetheless, to avoid having too late deliveries, constraints (8) are used.

4.2.2 Constraints

To simplify the interpretation of the model, nesting constraints and scheduling constraints are presented separately. Still, it is crucial to adequately link these two subproblems to tackle them from an integrated perspective, and thus, relationship constraints are also presented. Additionally, to break symmetries and obtain a more efficient model, symmetry-breaking constraints (SBC) were developed.

Nesting Constraints

$$X_{id} = 1 \Rightarrow \sum_{d' \in NFP^{d}_{i\,i'}} X_{i'd'} = 0, \forall i, i' \in I, i \neq i', d \in IFP_i \tag{2}$$

$$\sum_{d \in IFP_i} X_{id} = 1, \forall i \in I \tag{3}$$

Nesting is done based on the dotted-board model with binary variables of Cherri et al. [10], which allows modeling irregular-shaped parts. Constraints (2) guarantees that if the positioning point of part i is placed in dot d, then the positioning point of part i' cannot be placed in the dots d' that belong to the subset $NFP^{d}_{i\,i'}$, otherwise the two parts overlap. By placing the positioning point of part i in a certain dot d, the part will belong to the build to which the dot belongs, i.e., $i \in h_d$. Constraints (3) ensures that the positioning point of each part i is placed in one dot d to guarantee that the part is produced. Note that $d' \in NFP^{d}_{i\,i'}$ only if $d' \in IFP_{i'}$.

Scheduling Constraints

$$alternative\left(Y_b, \{N_{bm}\}\right), \forall b \in B, m \in M \tag{4}$$

$$noOverlap\left(S_m\right), \forall m \in M \tag{5}$$

The scheduling constraints involve the built-in functions of the CP optimizer $alternative()$ and $noOverlap()$. In constraints (4) a machine m from the set of available machines is chosen to process build b. Then, constraints (5) are responsible for defining the sequence of the different builds b assigned to machine m, guaranteeing that only one build is produced at a time.

Relationship Constraints

$$sizeOf\left(Y_{h_d}\right) \geq p_i \times X_{id}, \forall i \in I, d \in IFP_i \tag{6}$$

$$startOf\left(Y_{h_d}\right) \geq r_i \times X_{id}, \forall i \in I, d \in IFP_i \tag{7}$$

$$X_{id} = 1 \Rightarrow MD \times T_i \geq \left(endOf\left(Y_{h_d}\right) - d_i\right), \forall i \in I, d \in IFP_i \tag{8}$$

Constraints (6) and (7) guarantee that if the positioning point of a certain part i is assigned to dot d, i.e., $X_{id} = 1$, then the build to which dot d belongs, has a processing time at least as long as the processing time of part i, and never starts before the release date of part i. The build to which dot d belongs is obtained by the parameter h_d. Constraints (8) are used to guarantee that if a certain part i assigned to build h_d finishes after its due date d_i, it is late, and thus $T_i = 1$. However, it guarantees that such delay is not superior to the maximum delay allowed, MD.

Symmetry-Breaking Constraints

$$presenceOf\,(Y_{b-1}) \geq presenceOf(Y_b), \forall b \in B, b \geq 2 \tag{9}$$

$$\max_{b \in B} endOf\left(N_{b,m-1}\right) \geq \max_{b \in B} endOf\left(N_{b,m}\right), \forall m \in M, m \geq 2 \tag{10}$$

$$endOf\left(N_{b,m}\right) \geq \max_{b' \in B, b' < b} endOf\left(N_{b',m}\right) \times presenceOf\left(N_{b,m}\right), \quad \forall b \in B, b \geq 2, m \in M \tag{11}$$

Three different SBC were introduced. Since builds are sequentially numerated, constraints (9) intend to guarantee that a specific build is only used if the previous one is already used. Constraints (10) are used because machines are identical, making it possible to swap the schedule of one machine with the schedule of the other without changing the global solution. Lastly, constraints (11) intend to do something similar to constraints (9); however, it is done on a machine level. As an example, processing build 2 (containing part 3 and 4) and then build 4 (containing part 1 and 6) on a certain machine, is the same as processing build 4 (containing part 3 and 4) and then build 2 (containing part 1 and 6) on that same machine. Thus, constraints (11) guarantee that only the first solution is available.

5 Experimental Study

In the previous section, we proposed a CP model that integrates nesting and scheduling decisions to minimize the total costs involved in machines' utilization and tardy deliveries. To further understand this model, we study different instances. The instances generation and results obtained are shown in the following subsections.

5.1 *Instance Generation*

Parts' data were generated randomly, guaranteeing that parts' dimensions were smaller than the building platform. Both irregular and regular parts were taken into account, and the corresponding subsets IFP_i and $NFP^{d}_{i\,i'}$ were constructed in a pre-processing step.

Each building platform is considered to be a square of 60 × 60 cm, divided into 25 dots equally distributed according to the dotted-board model. The model was coded in IBM ILOG CPLEX Optimization Studio 20.1.0.0 using CP Optimizer, and all runs were done on a workstation with an Intel®Core i7-6800K 3.40GHz, and 64GB of RAM. No time limit was defined. Additionally, all problem instances considered two AM machines.

5.2 Results Analysis

We now examine the main results of the model. The first insight is related to the definition of the total number of builds n_b. Figure 3 shows the results obtained to produce 15 parts depending on the number of builds n_b, demonstrating the impact of this parameter in the model resolution. The instances are equal, except for the total number of builds n_b. As can be seen, a value of n_b too small can lead to infeasible solutions ($n_b = 3$), since there are not enough builds to nest all parts. However, a value of n_b too high can lead to a much higher computational time. This happens because more builds means more dots, which in turn increases the number of binary variables X_{id}. For $n_b = 5$ and $n_b = 6$, the solution remains the same as for $n_b = 4$, i.e., the optimal solution for $n_b = 5$ and for $n_b = 6$ uses 4 builds, leaving 1 and 2 builds free, respectively, but the computational time required is much higher.

In instances for producing 5 parts, results indicated that choosing a number of builds n_b too small can also lead to suboptimal solutions, since considering $n_b = 3$ to produce 5 parts leads to a better solution than the one obtained for $n_b = 2$. This means that the smaller number of builds required for a feasible solution is not necessarily the optimal solution, as happens in the case where 15 parts are produced.

Another conclusion is that SBC are efficient, particularly for the cases where the total number of builds n_b is defined higher than necessary. This effect can be seen in Fig. 4. The lower line demonstrates that as SBCs are added to the same instance, the necessary computational time decreases, with a difference between no SBCs and all SBCs of 25%. Here, the value of n_b is equal to the number of builds used by the optimal solution to produce 15 parts. However, looking at the upper line, where the instance considered has a value of n_b higher than the number of builds used by the optimal solution, it is clear that SBCs have a more significant effect. In these instances, SBCs led to a reduction of the computational time of approximately 55%.

After understanding the impact caused by the definition of n_b, as well as the impact of SBCs, it is essential to understand how the model reacts to the dimension of the instances. To do that, different instances were analyzed. The results indicated that the model can quickly obtain the optimal solution for a number of parts up to around 15. However, with the increase of the number of parts, the computational time required to obtain the optimal solution grows significantly. Nonetheless, the increase in the

Fig. 3 Results of producing 15 parts with different number of builds

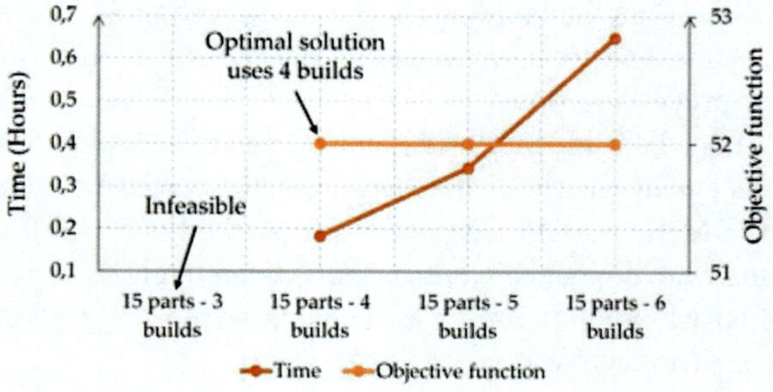

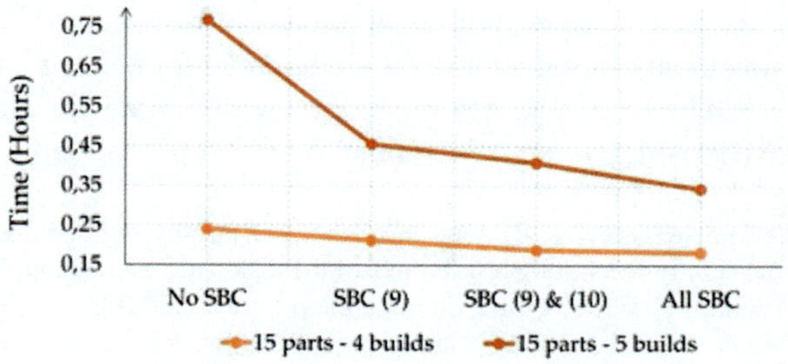

Fig. 4 Results of producing 15 parts with different number of builds

computational time is more related to proving optimality than finding the optimal solution. The model can find the optimal solution relatively fast once it reads and processes all the input data, but requires a significant amount of time to prove the optimality of the solution. As an example, when producing 20 parts with 5 builds, the model only requires approximately 3 min to find the optimal solution once it reads all the input data, but it requires an additional 384 min to prove optimality.

6 Conclusions and Future Research

We consider the simultaneous nesting and scheduling problem of identical parallel AM machines. The optimal allocation of parts into machines aims to minimize the total costs related to operating costs and tardy deliveries. Since the process of grouping parts into builds is typically simplified in the scarce literature about AM scheduling, one of the objectives was to use a nesting method as close to reality as possible able deal with irregular-shaped parts. To do that, the dotted-board model with binary variables for the irregular bin-packing problem of Cherri et al. [10] was adapted.

According to the results obtained, the definition of the parameter number of builds n_b is critical. Defining a value of n_b too low can lead to infeasible solutions, while a value of n_b too high can lead to much higher computational times without improving the solution. The addition of SBCs to the model, though, can efficiently reduce the negative effects of a wrong definition of this parameter. Additionally, the model has the potential to be used for AM scheduling, particularly for short-term planning, since to schedule the production of more than around 15 parts, computational times significantly increase. Nonetheless, such an increase is essentially related to proving optimality and not to finding the optimal solution.

This work is the first contribution to the literature regarding AM scheduling that uses CP as a modeling approach, as well as the first that integrates the dotted-board model in a scheduling problem, since up until now the dotted-board model has been exclusively used in nesting problems. Additionally, it is the first that considers an exact approach to the AM scheduling problem without simplifying nesting.

In future studies, two main paths will be followed. The first one is related to the implementation of a more efficient model for nesting, also proposed by Cherri et al. [10]. The second path is related to the possible hybridization of the CP model with a MIP model to reduce the time to prove optimality.

Acknowledgements This research is supported by the doctoral Grant SFRH/BD/151417/2021 financed by the Portuguese Foundation for Science and Technology (FCT), under MIT Portugal Program, by FEDER funds through the program COMPETE - Programa Operacional Factores de Competitividade; and by national funds through FCT under the project UID/EMS/00285/2020; and by the Operational Programme for Competitivity and Internationalization of Portugal 2020 Partnership Agreement (PRODUTECH4S&C), grant number POCI-01-0247-FEDER-046102.

References

1. Thürer, M., Huang, Y., Stevenson M.: Workload control in additive manufacturing shops where post-processing is a constraint: an assessment by simulation. Int. J. Prod. Res. 1–19 (2020). https://doi.org/10.1080/00207543.2020.1761038
2. Oh, Y., Witherell, P., Lu, Y., Sprock, T.: Nesting and scheduling problems for additive manufacturing: a taxonomy and review. Addit. Manuf. **36**, 101492 (2020). https://doi.org/10.1016/j.addma.2020.101492
3. Che, Y., Hu, K., Zhang, Z., Lim, A.: Machine scheduling with orientation selection and two-dimensional packing for additive manufacturing. Comput. Oper. Res. **130**, 105245 (2021). https://doi.org/10.1016/j.cor.2021.105245
4. Zhang, J., Yao, X., Li, Y.: Improved evolutionary algorithm for parallel batch processing machine scheduling in additive manufacturing. Int. J. Prod. Res. **58**, 2263–2282 (2020). https://doi.org/10.1080/00207543.2019.1617447
5. Li, X., Zhang, K.: Single batch processing machine scheduling with two-dimensional bin packing constraints. Int. J. Prod. Econ. **196**, 113–121 (2018). https://doi.org/10.1016/j.ijpe.2017.11.015
6. Uzsoy, R.: Scheduling a single batch processing machine with non-identical job sizes. Int. J. Prod. Res. **32**, 1615–1635 (1994). https://doi.org/10.1080/00207549408957026
7. Arroyo, J.E.C., Leung, J.Y.T.: Scheduling unrelated parallel batch processing machines with non-identical job sizes and unequal ready times. Comput. Oper. Res. **78**, 117–128 (2017). https://doi.org/10.1016/j.cor.2016.08.015
8. Ham, A., Fowler, J.W., Cakici, E.: Constraint programming approach for scheduling jobs with release times, non-identical sizes, and incompatible families on parallel batching machines. IEEE Trans. Semicond. Manuf. **30**, 500–507 (2017). https://doi.org/10.1109/TSM.2017.2740340
9. Toledo, F.M.B., Carravilla, M.A., Ribeiro, C., et al.: The dotted-board model: a new MIP model for nesting irregular shapes. Int. J. Prod. Econ. **145**, 478–487 (2013). https://doi.org/10.1016/j.ijpe.2013.04.009
10. Cherri, L.H., Carravilla, M.A., Ribeiro, C., Toledo, F.M.B.: Optimality in nesting problems: new constraint programming models and a new global constraint for non-overlap. Oper. Res. Perspect. **6**, 100125 (2019). https://doi.org/10.1016/j.orp.2019.100125
11. Kucukkoc, I., Li, Q., Zhang, D.Z.: Increasing the utilisation of additive manufacturing and 3D printing machines considering order delivery times. In: Nineteenth International Working Seminar on Production Economics (2016)
12. Li, Q., Kucukkoc, I., Zhang, D.Z.: Production planning in additive manufacturing and 3D printing. Comput. Oper. Res. **83**, 1339–1351 (2017). https://doi.org/10.1016/j.cor.2017.01.013

Developing a System for Sectorization: An Overview

Elif Göksu Öztürk, Filipe Soares de Sousa, Maria Margarida Lima, Pedro Filipe Rocha, Ana Maria Rodrigues, José Soeiro Ferreira, Ana Catarina Nunes, Isabel Cristina Lopes, and Cristina Teles Oliveira

Abstract Sectorization is the partition of a set or region into smaller parts, taking into account certain objectives. Sectorization problems appear in real-life situations, such as school or health districting, logistic planning, maintenance operations or transportation. The diversity of applications, the complexity of the problems and the difficulty in finding good solutions warrant sectorization as a relevant research area. Decision Support Systems (DSS) are computerised information systems that may provide quick solutions to decision-makers and researchers and allow for observing differences between various scenarios. The paper is an overview of the development of a DSS for Sectorization, its extent, architecture, implementation steps and benefits. It constitutes a quite general system, for it handles various types of problems, which the authors grouped as (i) basic sectorization problems; (ii) sectorization problems with service centres; (iii) re-sectorization problems; and (iv) dynamic sectorization problems. The new DSS is expected to facilitate the resolution of various practitioners' problems and support researchers, academics and students in sectorization.

Keywords Decision support systems · Sectorization · Evolutionary algorithms

---

E. Göksu Öztürk (✉) · P. Filipe Rocha · A. Maria Rodrigues · J. Soeiro Ferreira
INESCTEC - Technology and Science, Porto, Portugal
e-mail: elif.ozturk@inesctec.pt

M. Margarida Lima · A. Maria Rodrigues · I. Cristina Lopes · C. Teles Oliveira
CEOS.PP, ISCAP, Polytechnic of Porto, Porto, Portugal

F. Soares de Sousa · A. Catarina Nunes
ISCTE - University Institute of Lisbon, Lisbon, Portugal

J. Soeiro Ferreira
FEUP - Faculty of Engineering, University of Porto, Porto, Portugal

E. Göksu Öztürk
FEP.UP - Faculty of Economics, University of Porto, Porto, Portugal

A. Catarina Nunes
CMAFcIO - Faculty of Sciences, University of Lisbon, Lisbon, Portugal

J. P. Almeida et al. (eds.), *Operational Research*, Springer Proceedings in Mathematics & Statistics 411,
https://doi.org/10.1007/978-3-031-20788-4_9

1 Introduction

Sectorization is a division of a large area, territory or network into smaller parts considering one or more objectives. Sectorization problems are diverse given the vast fields of application. Different applications arise due to the need for several real-life dilemmas. Examples include designing political districts, sales territories, schools, health and policing zones, forest planning, municipal waste collection or street cleaning zones and maintenance operations. Solution procedures are challenging given the wide range of applications and problem complexities.

A good sectorization creates economic, social, and financial benefits. However, the definition of a good sectorization can be very subjective and can change from one Decision Maker (DM) to another depending on different 'what if' scenarios (i.e. restrictions and objectives). A Decision Support System (DSS) designed for Sectorization can support analysing and resolving different problems under diverse conditions.

In general terms, DSS refers to an information system that provides computer help to DMs or its users to solve their specific problems. DSS are strong tools since they allow multiple users, supply data and provide analytics. Moreover, DSS are useful for policy-making. DSS users can easily prove their 'what if' scenarios using these tools to observe the differences in the solutions, compare, and decide. Thus, DSS play a vital role in increasing time and cost efficiency, providing better performance management, and empowering users' decisions even if the problem is not well-defined or immature [26].

The paper's main contribution is the brief description of a new DSS dedicated to Sectorization (D3S) problems. D3S deals with different multiple criteria problems grouped as (i) basic sectorization problems; (ii) sectorization problems with service centres; (iii) re-sectorization problems; and (iv) dynamic sectorization problems. This platform is designed in the format of a website with a user-friendly outline. The authors are not aware of any DSS for sectorization, which is so comprehensive. Thus, it is expected that the new D3S will help researchers in the sectorization field and contribute to dealing with real cases involving different criteria and scenarios.

The remainder of the paper is structured as follows. Section 2 provides a literature review on application fields of sectorization problems and DSS that aim to assist the solution procedure of these problems. Section 3 includes detailed information about D3S and the four steps that the user should follow to use the system. Solution methods used in D3S to solve sectorization problems are briefly presented in Sect. 4. Finally, Sect. 5 concludes with a discussion on future work.

2 Literature Review

The current section is a short review of the literature concerning Sectorization and DSS.

As mentioned, Sectorization problems appear in various real-world settings under different configurations. For instance, political districting is one of the oldest application fields. Political districting problems aim to divide the territory neutrally and avoid gerrymandering. It is possible to see several applications of political districting considering various criteria such as equilibrium, compactness, contiguity, or community interest in the literature [3, 18]. Most of the time, political districting is subject to resectorization problems to adapt the current solution regarding the updates in the territory. Bozkaya et al. [4] built a spatial DSS as a built-in within ArcGIS called "DistrictBuilder". They used a tabu search algorithm and designed the system properly for several objectives. The authors tested their DSS to resectorize the political districts of Edmonton City in Canada.

Moreover, the design of sales and service territories is another field of sectorization. These problems include strategic logistic planning such as determining the locations of factory depots, distribution centres or sale facilities and designing service territories in terms of customer and/or technical facilities and maintenance operations [16, 17, 20]. Routing is usually a critical stage when the sectorization process is completed. Designing compact sectors for the salespersons or customer service workers to travel between clients most efficiently requires good sectorization first. Thus, most of the time, the solutions should guarantee aspects like connectivity within the sectors, balance in work hours, travel time or distance. It is common to consider predefined service centres or plans to determine some during the solution procedure in these problems. Moynihan et al. [22] developed a micro-computer-based DSS for logistic/distribution network planning. They provided test alternatives for the DMs to observe the possible solutions under different scenarios and objectives. The solution methods implemented in the system are based on simulation and heuristics. Another effort was taken place by Noorian and Murphy [24] on territory design problems. The authors aimed to create optimally balanced sectors by simultaneously respecting region restrictions and multiple objectives. The solution method implemented was a Genetic Algorithm (GA) based on graph theory. This web-based DSS was shared with the user through the intermediate platform SaaS.

Furthermore, the distribution of energy and microcells are also relevant fields of sectorization. It is important to distribute the mobile network, internet and electricity in the most similar and fast way since they are limited and essential resources for human life [5, 7, 27]. In the solution procedure of these problems, microcell and energy towers can be viewed as service centres/facilities. Bergey et al. [1] built a DSS to support electric power districting (EPD). They create an EPD problem solver as a Microsoft Excel built-in. They implemented GA with a configuration that automatically guarantees the contiguity in the distribution network. This system allowed the DMs to visualise the frontier solutions and pick the most convenient one.

Other common fields to be studied are related to the problems of waste collection and water distribution. These applications play a fundamental role in improving cities' infrastructure and increasing living standards [21, 28]. Pan et al. [25] developed a DSS to operate Milan, Italy's daily and long-term water distribution network. The authors aim to use this DSS for both pump scheduling and the sectorization of the system to make benefits in the daily and eventually long term operations in Milan city. They developed a web-based DSS with two metaheuristic methods as a solver, namely, the Non-dominated Sorting Genetic Algorithm (NSGA-II) and Archive-based Micro Genetic Algorithm (AMGA2).

Additionally, sectorization applications on forest planning aim to improve the ecological (by protecting the wildlife), economic (by reinforcing sustainability) and social (by promoting leisure and tourism) aspects of forest management [23, 27, 36]. Wilksröm et al. [11] built a DSS on forest planning. The system supported the stand, forest and regional analysis and planning. The platform was built in the format of a desktop app. They followed Analytical Hierarchy Process (AHP) to scale multiple criteria when it is the case.

A narrower field of sectorization necessary during winter in many cities is snow removal or disposal. Heavy snow shall negatively affect living standards by blocking traffic and diminishing mobility from one place to another. This situation may have social and economic consequences [19]. Thus, designing the cities and leading the municipal service to the necessary points in the most balanced and fastest way are the goals of this type of problem. Depending on the weather conditions or the countries' infrastructure, these problems may be considered dynamic or subject to resectorization. Labelle et al. [19] developed a DSS to sectorize the cities for snow removal quickly and efficiently. They used MapInfo GIS software to build their DSS, which was beneficial due to the possibility of instant interaction of the DM with the provided solution. This way, they aimed to give the most convenient solution to the DMs. They implemented the DSS in Montreal to observe the performance of their system.

Additionally, problems on schooling, police, health districting and transportation are common problems beneath sectorization [2, 6, 12, 38]. School districting carries dynamic attributes. It is important to assign the same pupil to the same school from one year to another while distributing the new pupils to the most convenient schools by considering the transportation costs and the institution capacity [2]. Health districting mostly provides caregiving services to the patients in a most balanced way. These problems can be considered dynamic since the potential changes in the unmet health needs of the patients.

Given the need to update school districting every year, it is possible to observe various efforts to build DSS on this specific sectorization field to make the decisions as quickly as possible. For instance, Ferland and Guenette [14] designed a microcomputer-based DSS. They used heuristic methods and offered solutions regarding contiguity while considering school capacity, population and number of students in different schools. The platform allowed modification of the solution if needed, besides the user integration.

Camacho-Collados and Liberatore [6] built a DSS, called Predictive Police Patrolling DSS (P^3-DSS) to construct powerful and balanced police patrolling sectorization to distribute police officers efficiently and decrease criminal actions.

Moreover, air-traffic services can be efficiently designed using sectorization methods. Essentially, airspace sectorization is one of the major application fields given its international and international importance in the line of transportation [9, 34]. Weigang et al. [37] built a distributed DSS for air traffic flow management. The meta-level control approach and reinforcement-learning algorithms aim to decrease the possible risks in air traffic flow to the minimum. Stamatopoulos et al. [33] developed a DSS for strategic airport planning. They distinguished the drivers of the airport as dynamic and stochastic. This way, they created a structural model which could interact with the different areas of the airfield by considering the specific need of that area.

Finally, we refer to two classical books, which can still contribute to developing model-driven DSS, especially if supplemented with new computerised and communication means [32, 35].

As is seen, there are various efforts to build DSS to solve different sectorization problems. Most DSS use Evolutionary Algorithms as solution methods. It is possible to use some of these systems for problems not intended for. However, the authors are unaware of any DSS dedicated to dealing with diversified sectorization problems. D3S was genuinely designed to represent such a generic system for sectorization practice and field literature.

3 System Design

As mentioned, the Decision System, which will be called D3S, is designed as a web-based DSS using Python base Web-framework Django since any device with a web browser can access it. It is a straightforward and user-friendly platform. Figure 1 shows the system architecture. The interaction with the system starts with the User Interface, through which the user's problems are proposed.

The model management module keeps the inputs, namely, user preferences, problem information, problem instance (i.e. data) in the database. It then processes this information and sends it to the adequate problem solver. The model management module includes several problem-solvers for different scenarios. The final solutions obtained are stored in the database and can be accessed again through the user interface. This process can be seen in the flow diagram in Fig. 2.

The remainder of this section provides detailed information about the four steps a user should follow to benefit from D3S.

Step 1: Select the Service

D3S has been developed to deal with sectorization. Figure 3 shows the services that D3S provides within four groups, introduced by the authors when considering the general attributes of different sectorization problems, as articulated in Sect. 2. These groups are: (i) basic sectorization problems that do not consider any predetermined service centres nor plan to determine service centres; (ii) sectorization problems with service centres that consider predetermined or fixed service centres or plans to determine one; (iii) redistricting problems for redesigning a solution that is already

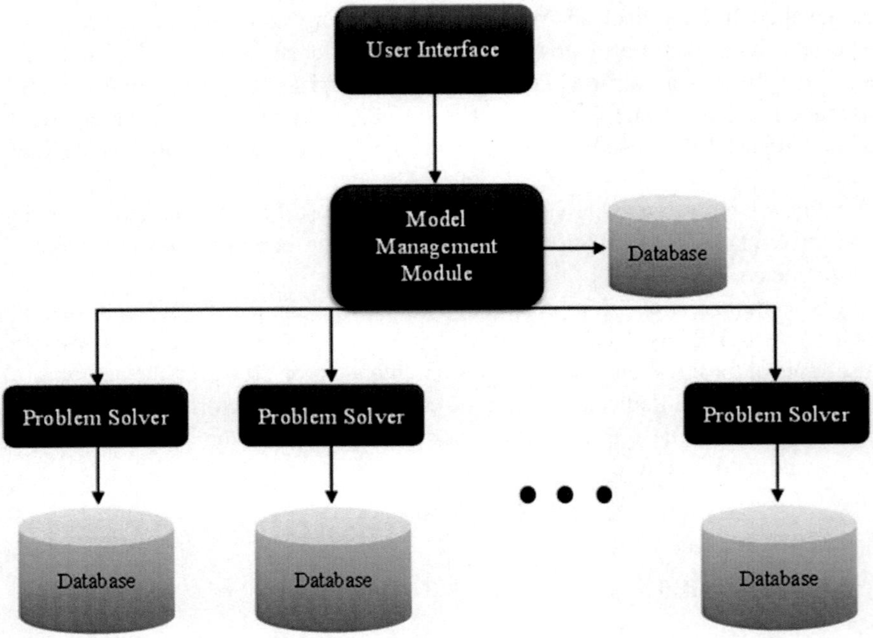

Fig. 1 System architecture

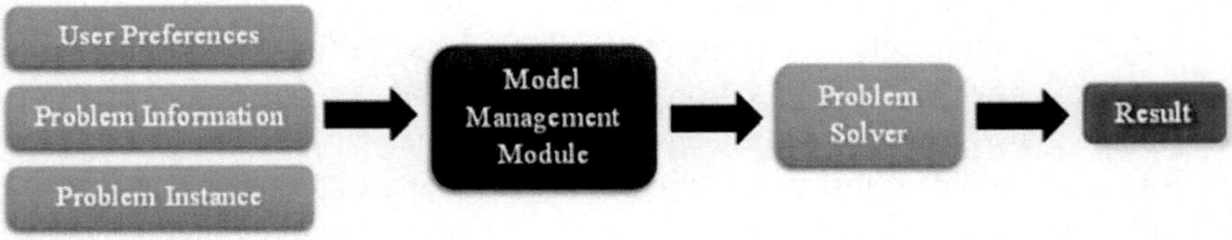

Fig. 2 Input-output diagram

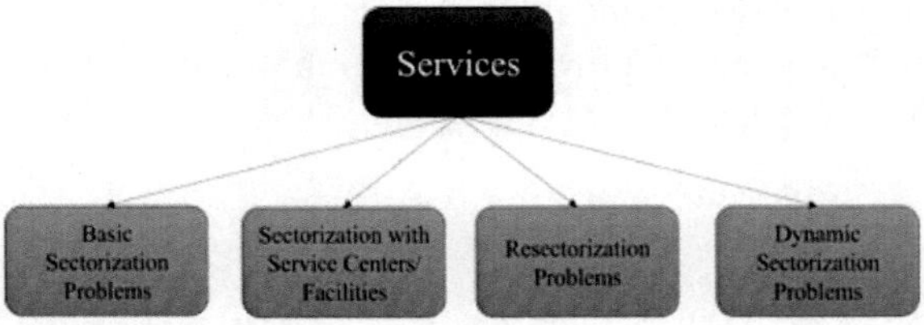

Fig. 3 Services provided by D3S

valid; (iv) dynamic sectorization problems that represent more than one point in time and allow changes on the instances over time.

As the initial step of D3S, the users are expected to select one of these services proper for their problem.

Step 2: Fill in the Survey

After users select the related service, they are required to respond to a more detailed survey to input the specific attributes of the problem. This survey is structured to understand the nature of the problem, the basic characterisation of the problem, the type of solution wanted, and the criteria and objectives considered. Through the questions in these categories, D3S identifies the problem in specific.

For instance, the nature of the problem includes questions regarding the distance considered and the characterisation of the basic units. More precisely, users may prefer different distance types between the basic units, such as Euclidean or other. *Basic Units* can be defined as points or geographic areas. Furthermore, whether the basic units are subject to a weighting scheme (e.g. valorisation, demand, or the number of customers) is also beneath the nature of the problem.

Moreover, the basic characterisation of the problem includes questions that aim to understand the features in more detail. This category contains questions regarding the number of sectors and their capacity or whether the user wants to receive routing within each sector. When the problem consists of service centres/facilities, the questions aim to understand whether there are predefined service centres or whether the users plan to define them. If there are predefined service centres, whether all will be used or whether a service centre will serve only one sector or more. Furthermore, when resectorization is in question, whether the user looks for a completely new solution or asks for a solution similar to the one that already exists. At last, in the case of dynamic problems, the questionnaire also collects information regarding how many time horizons the user considers and the percentage changes in the instances that could occur from one moment to another.

The solution type aims to understand whether the user expects to obtain single or multiple solutions. The answer to this question directly affects the solution method that D3S uses.

Finally, the category called criteria and objectives collects the information regarding the user's objectives, such as balance, connectivity, density, and time distance minimisation. Moreover, users can give information about if there are natural boundaries to be respected or, on the contrary, if some basic units need to be in the same sector.

As is seen, only the category regarding the basic characterisation of the problem changes from service to service. The rest of the questionnaire remains the same.

These surveys are essential since the answers of the users have a direct effect on the optimisation process. Every service has its specific query set for problem identification.

Step 3: Upload Data

The data uploading phase starts following the survey submission. The users are expected to present their instances according to a given structure of the system. The data expected by the system require some specific information to execute the optimisation. The required information shows some diversity from one service to another given the different nature of the services.

The information, which constitutes latitude, longitude, weights (i.e. quantity or demand), links, boundaries, and neighbours, is fixed in the data for all services. Besides, service centres and their capacities are also necessary when sectorization with service centres are considered. Furthermore, information regarding the old solution is expected in resectorization problems.

To avoid complications in this phase, given the distinctions in the expected data structures, the system first provides a downloadable template with detailed explanations about the data structure. The template can be downloaded and filled with the instances in the proper format.

Moreover, the system provides instances to test the D3S for valued objectives if the users do not have data. These instances can be found through the following link: https://drive.inesctec.pt/s/NS47qnZEmYPwEQP.

Submission of the data activates the model management module. The optimisation is executed according to the answers to the survey using the uploaded data.

Step 4: Get the Results and Visualise

After data submission, D3S takes the users to a 'Results' page to see all their submissions. Each submission is a link to a page that contains a detailed summary of that specific submission. In these summaries, users can observe the type of service they benefited from, their answers to each survey question, their data, computation time, and the result(s) provided by the D3S.

The users can observe the fitness scores of each solution before selecting. This score is a single value containing the overall performance (case of single solution) or multiple values that show the solution's fitness for different objectives separately (case of multiple solutions). Besides that, it is also possible to visualise the solution(s), which will facilitate any final user decision phase. Figure 4 shows a small print screen from D3S that represents the process explained in this section.

The user can see all the submissions with the dates. Each submission links and leads the user to the submission summary page with detailed information and the results provided. In the example presented in Fig. 4, we observe multiple solutions with three objectives: equilibrium, compactness, and contiguity. By clicking on the "Solution N", one can visualise a map of the results provided by D3S.

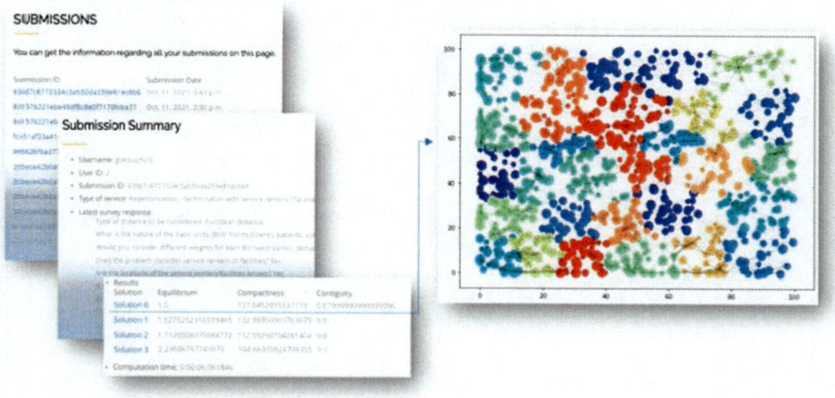

Fig. 4 Results provided by D3S

4 Solution Methods

D3S is a platform that provides solutions according to different preferences. The solution methods that the D3S uses are based on Evolutionary Algorithms. As mentioned in Sect. 3, choosing the preferred solution method (i.e. single or multiple) is possible.

D3S follows a GA when the user asks for a single solution. GA, a well-known algorithm proposed by Goldberg and Holland [15], evaluates the solutions according to their performance on the fitness function through generations. The fitness of a solution represents the adequacy of a solution to the problem. It is also a commonly used and powerful algorithm to solve sectorization problems [5, 10, 24].

If more than one objective is in question, the fitness function must be built as a weighted composite single objective function in GA. Establishing this composite equation requires normalisation of the objectives with different measurement units and weighing processes. In D3S, AHP, presented by Saaty [29–31], is followed to build the weighting scheme based on the preferences of the DMs among the objectives. These preferences are detected by ordering them within the pairwise comparison scale, which contains elements from 1 to 9, and higher values show stronger importance of one objective over another. The DMs are expected to compare each objective unilaterally with any other.

On the other hand, the NSGA-II is used to solve the optimisation problem when the user demands multiple solutions. In that case, D3S provides Pareto frontier solutions to the user's decision.

NSGA-II is one of the most used multi-objective optimisation methods presented by Deb et al. [8]. In this method, Pareto frontier solutions are selected according to their performance in the solution space. A solution is superior if and only if that solution is better than the other solution in at least one objective while not being

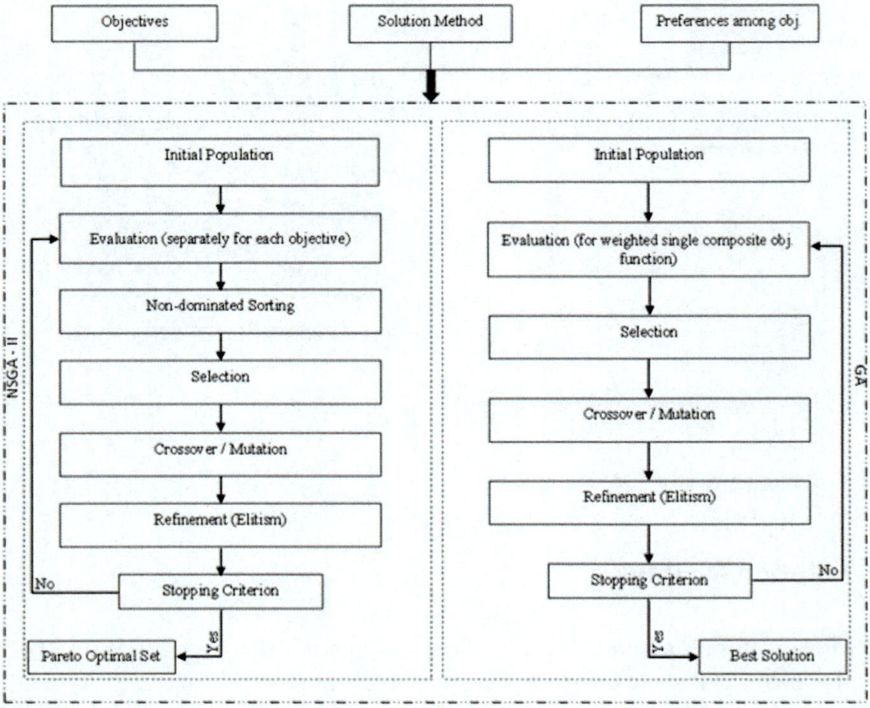

Fig. 5 Schematic of the solution procedures

worse in the other objectives. Moreover, NSGA-II is selected to be implemented in the D3S as an adequate approach since several authors use it in sectorization and similar problems [13, 39, 40]. Furthermore, Pan et al. [25] applied NSGA-II as one of the solution methods in their web-based DSS on the water distribution network of Milan.

The necessary steps to complete a generation (i.e. iteration) are similar in GA and NSGA-II. This situation is visible in Fig. 5. As is seen, the main difference between these methods is the evaluation process of the solutions. Thus, besides the selected solution type, objectives and user preferences are vital inputs for the solution methods used in D3S.

Finally, the D3S platform uses a Greedy Algorithm for routing if the user asks for it. The routing moves to the closest basic unit in this algorithm. The Euclidean distances are considered unless the user presents data regarding the type of distances that s/he would like to take into account.

5 Conclusion and Future Work

Sectorization refers to dissolving a whole into subsets by respecting some objectives. Diversity in the application areas in real-life problems makes sectorization a very relevant field of study. For instance, design of sales or service territories, maintenance operations, health and school districting, police patrolling or transportation are some of the applications that may directly affect countries, economies, or human life itself.

Sectorization solution procedures can be difficult due to uncertainties in their characterisation, objectives' choice, and the frequent combinatorial nature of the associated problems. DSS are computer-based information systems that can provide solutions rapidly to test, adjust, and decide.

The current work presented D3S, a new web-based DSS designed to solve various sectorization problems, organised in the four revealed groups. D3S was developed to solve sectorization problems arranged within four main groups: basic sectorization, sectorization with service centres, resectorization and dynamic sectorization. Solution methods integrated into the system are NSGA-II or GA with AHP to build fitness functions. They will be used depending on the type of solution desired by users. One of the D3S advantages is its flexibility in terms of objectives and restrictions and the applicability to various sectorization problems.

D3S will evolve by improving existing algorithms, integrating new optimisation algorithms and dealing with more objectives. The system will soon be publicly available at www.sectorization.pt We hope that the D3S will contribute to the resolution of various practitioners' sectorization problems and assist researchers, academics and students.

A short video of D3S can be found at https://www.researchgate.net/project/StoSS-Sectorization-to-Simplify-and-Solve/update/61c375edf5675b211b18c582.

Acknowledgements This work is financed by the ERDF - European Regional Development Fund through the Operational Programme for Competitiveness and Internationalisation - COMPETE 2020 Programme and by National Funds through the Portuguese funding agency, FCT - Fundação para a Ciência e a Tecnologia within project "POCI-01-0145-FEDER-031671".

References

1. Bergey, P.K., Ragsdale, C.T., Hoskote, M.: A decision support system for the electrical power districting problem. Decis. Support Syst. **36**(1), 1–17 (2003)
2. Bouzarth, E.L., Forrester, R., Hutson, K.R., Reddoch, L.: Assigning students to schools to minimize both transportation costs and socioeconomic variation between schools. Socioecon. Plann. Sci. **64**, 1–8 (2018)
3. Bozkaya, B., Erkut, E., Laporte, G.: A tabu search heuristic and adaptive memory procedure for political districting. Eur. J. Oper. Res. **144**(1), 12–26 (2003)
4. Bozkaya, B., Erkut, E., Haight, D., Laporte, G.: Designing new electoral districts for the city of edmonton. Interfaces **41**, 534–547 (2011)
5. Brown, E.C., Vroblefski, M.: A grouping genetic algorithm for the microcell sectorization problem. Eng. Appl. Artif. Intell. **17**(6), 589–598 (2004)

6. Camacho-Collados, M., Liberatore, F., Angulo, J.M.: A multi-criteria police districting problem for the efficient and effective design of patrol sector. Eur. J. Oper. Res. **246**(2), 674–684 (2015)
7. De Assis, L.S., Franca, P.M., Usberti, F.L.: A redistricting problem applied to meter reading in power distribution networks. Comput. Oper. Res. **41**, 65–75 (2014)
8. Deb, K., Pratap, A., Agarwal, S., Meyarivan, T.: A fast and elitist multiobjective genetic algorithm: Nsga-ii. IEEE Trans. Evol. Comput. **6**(2), 182–197 (2002)
9. Degtyarev, O., Minaenko, V., Orekhov, M.: Solution of sectorization problems for an air traffic control area. i. basic principles and questions of airspace sectorization and its formalization as an optimization problem. J. Comput. Syst. Sci. Int. **48**(3), 384–400 (2009)
10. Di Nardo, A., Di Natale, M., Santonastaso, G.F., Tzatchkov, V.G., Alcocer-Yamanaka, V.H.: Water network sectorization based on a genetic algorithm and minimum dissipated power paths. Water Sci. Technol.: Water Supply **13**(4), 951–957 (2013)
11. Edenius, L., Elfving, B., Eriksson, L.O., Sonesson, J., Wallerman, J., Waller, C., et al.: The heureka forestry decision support system: an overview. Math. Comput. For. Nat. Resour. Sci. **3**(2) (2011)
12. Farughi, H., Mostafayi, S., Arkat, J.: Healthcare districting optimization using gray wolf optimizer and ant lion optimizer algorithms (case study: South khorasan healthcare system in Iran). J. Optim. Ind. Eng. **12**(1), 119–131 (2019)
13. Farughi, H., Tavana, M., Mostafayi, S., Santos Arteaga, F.J.: A novel optimization model for designing compact, balanced, and contiguous healthcare districts. J. Oper. Res. Soc. **71**(11), 1740–1759 (2020)
14. Ferland, J.A., Guénette, G.: Decision support system for the school districting problem. Oper. Res. **38**(1), 15–21 (1990)
15. Goldberg, D.E., Holland, J.H.: Genetic algorithms and machine learning (1988)
16. Jahuira, C., Cuadros-Vargas, E.: Solving the tsp by mixing gas with minimal spanning tree. Sociedad Peruana de Computacion, II-3, pp. 123–133 (2003)
17. Kalcsics, J., Nickel, S., Schröder, M.: Towards a unified territorial design approach-applications, algorithms and GIS integration. TOP **13**(1), 1–56 (2005)
18. Kim, M.J.: Multiobjective spanning tree based optimization model to political redistricting. Spat. Inf. Res. **26**(3), 317–325 (2018)
19. Labelle, A., Langevin, A., Campbell, J.F.: Sector design for snow removal and disposal in urban areas. Socioecon. Plann. Sci. **36**(3), 183–202 (2002)
20. Lei, H., Wang, R., Laporte, G.: Solving a multi-objective dynamic stochastic districting and routing problem with a co-evolutionary algorithm. Comput. Oper. Res. **67**, 12–24 (2016)
21. Mourão, M.C., Nunes, A.C., Prins, C.: Heuristic methods for the sectoring arc routing problem. Eur. J. Oper. Res. **196**(3), 856–868 (2009)
22. Moynihan, G.P., Raj, P.S., Sterling, J.U., Nichols, W.G.: Decision support system for strategic logistics planning. Comput. Ind. **26**(1), 75–84 (1995)
23. Murray, A.T.: Spatial restrictions in harvest scheduling. For. Sci. **45**(1), 45–52 (1999)
24. Noorian, S.S., Murphy, C.E.: Balanced allocation of multi-criteria geographic areas by a genetic algorithm. In: International Cartographic Conference, pp. 417–433. Springer, Berlin (2017)
25. Pan, Q., Castro-Gama, M.E., Jonoski, A., Popescu, I.: Decision support system for daily and long term operations of the system of Milan, Italy. Procedia Eng. **154**, 58–61 (2016)
26. Ramachandra, T., George, V., Vamsee, K.S., Purnima, G.: Decision support system for regional electricity planning. Energ. Educ. Sci. Technol. **17**(1/2), 7 (2006)
27. Richards, E.W., Gunn, E.A.: Tabu search design for difficult forest management optimization problems. Can. J. For. Res. **33**(6), 1126–1133 (2003)
28. Ríos-Mercado, R.Z., Bard, J.F.: An exact algorithm for designing optimal districts in the collection of waste electric and electronic equipment through an improved reformulation. Eur. J. Oper. Res. **276**(1), 259–271 (2019)
29. Saaty, T.L.: A scaling method for priorities in hierarchical structures. J. Math. Psychol. **15**(3), 234–281 (1977)
30. Saaty, T.L.: The Analytical Hierarchy Process: Planning, Priority Setting, Resource Allocation. McGrawrHill International Book Co, London, England (1980)

31. Saaty, T.L.: Decision making with the analytic hierarchy process. Int. J. Serv. Sci. **1**(1), 83–98 (2008)
32. Sprague Jr., R.H., Carlson, E.D.: Building effective decision support systems. Prentice Hall Professional Technical Reference (1982)
33. Stamatopoulos, M.A., Zografos, K.G., Odoni, A.R.: A decision support system for airport strategic planning. Transp. Res. Part C: Emerg. Technol. **12**(2), 91–117 (2004)
34. Tang, J., Alam, S., Lokan, C., Abbass, H.A.: A multi-objective approach for dynamic airspace sectorization using agent based and geometric models. Transp. Res. Part C: Emerg. Technol. **21**(1), 89–121 (2012)
35. Turban, E., Aronson, J.E., Liang, T.P.: Decision Support Systems and Intelligent Systems Edisi 7 Jilid 1. Andi, Yogyakarta (2005)
36. Varma, V.K., Ferguson, I., Wild, I.: Decision support system for the sustainable forest management. For. Ecol. Manage. **128**(1–2), 49–55 (2000)
37. Weigang, L., de Souza, B.B., Crespo, A.M.F., Alves, D.P.: Decision support system in tactical air traffic flow management for air traffic flow controllers. J. Air Transp. Manag. **14**(6), 329–336 (2008)
38. Yanık, S., Kalcsics, J., Nickel, S., Bozkaya, B.: A multi-period multi-criteria districting problem applied to primary care scheme with gradual assignment. Int. Trans. Oper. Res. **26**(5), 1676–1697 (2019)
39. Zhang, K., Yan, H., Zeng, H., Xin, K., Tao, T.: A practical multi-objective optimization sectorization method for water distribution network. Sci. Total Environ. **656**, 1401–1412 (2019)
40. Zou, X., Cheng, P., An, B., Song, J.: Sectorization and configuration transition in airspace design. Math. Probl. Eng. (2016)

New Models for Finding K Short and Dissimilar Paths

Marta Pascoal, Maria Teresa Godinho, and Ali Moghanni

Abstract We address the problem of finding sets of K paths, $K \in \mathbb{N}$, which simultaneously considers two criteria: the minimization of the total paths' cost and the maximization of their dissimilarity. The purpose of these objectives is to find cheap solutions fairly different from one another, which are relevant considerations in applications that range from hazardous materials transportation to cash collection, where aspects like the safety or the reliability of the solutions are concerns.Two approaches are used to measure the dissimilarity of a set of paths: the extent of the overlap of the paths, in terms of the number of times that each arc appears in more than one of them; and the number of times that the arcs shared by two or more paths appear in that solution. The bi-objective problems resulting from each of these approaches are modeled in terms of integer linear programs, and an ε-constraint method is then designed to solve them. Computational results are presented for the two approaches in terms of the time efficiency, the quality of the sets of solutions obtained, and the dissimilarity of the efficient solutions.

Keywords K alternative paths · Cost · Dissimilarity · Integer Linear Programming formulations · Bi-objective optimization

M. Pascoal (✉) · A. Moghanni
Department of Mathematics, University of Coimbra, CMUC, 3001-501 Coimbra, Portugal
e-mail: marta@mat.uc.pt

M. Pascoal
Institute for Systems Engineering and Computers – Coimbra, rua Sílvio Lima, Pólo II, 3030-290 Coimbra, Portugal

Dipartimento di Elettronica, Informazione e Bioingegneria, Politecnico di Milano, Piazza Leonardo da Vinci, 32, Milan 20133, Italy

M. T. Godinho
Department of Mathematics and Physical Sciences, Polytechnic Institute of Beja, Campus do Instituto Politécnico de Beja, rua Pedro Soares, 7800-295 Beja, Portugal

CMAFcIO, Faculdade de Ciências da Universidade de Lisboa, Campo Grande, 1749-016 Lisbon, Portugal

J. P. Almeida et al. (eds.), *Operational Research*, Springer Proceedings in Mathematics & Statistics 411,
https://doi.org/10.1007/978-3-031-20788-4_10

1 Introduction

The problem of finding K paths in a network is of interest in many applications. In most cases, it is convenient to impose, or to value, that the number of resources they share is minimized, as the K paths often work as alternatives/backups to one another in case of a failure in the network. When looking for paths with those characteristics, we say that we aim at finding dissimilar paths. Figure 1 illustrates the concept of dissimilarity: the solid and the dotted paths share less resources, and thus are more dissimilar, than the dashed and dotted ones. Likewise, it is important that the alternative paths fulfill other requirements of the problem in consideration, namely keeping the solutions at a low cost. Therefore, a bi-objective version of the K paths problem, in which it is intended the minimization of the total cost of the K paths and the maximization of its dissimilarity arises in many practical applications—hazardous materials routing, money collection or telecommunications are examples of such applications. In the following, we designate the aforementioned problem as the K short and dissimilar paths problem.

The problem of finding sets of paths simultaneously short and dissimilar is approached by Dell'Olmo et al. [5] in the context of an application to hazardous material transportation. In this work, a large set of non-dominated paths with respect to a number of parameters that characterize the risk of the paths is generated by means of a multi-objective shortest path algorithm, followed by the resolution of a p-dispersion problem for selecting the paths according to their dissimilarity. Instead, Martí et al. [11] tackle directly the biobjective problem where the average cost is minimized and the dissimilarity of paths is maximized, while introducing a greedy randomized adaptive search procedure. The authors first review other works that maximize the paths dissimilarity, and then extend them to also take into consideration the paths cost [1, 3, 9, 10]. An application involving both spatial and time dissimilarities is presented in [15].

The K short and dissimilar paths problem was recently addressed in [14]. In this study, two versions of this bi-objective problem that differ in the approach used to model the dissimilarity of the paths are formulated: one of the versions focuses the minimization of the number of repeated arcs; the other one focuses the minimization of the total number of arc reuses. Then, the addition of a given extra constraint to the two models is considered. The resulting models are tested in two frameworks, corresponding to different variants of the ε-constraint method, according to how the parameter ε is updated. Recently, the same authors introduced complementary approaches to find dissimilar paths by means of single-commodity flow integer linear programming (ILP) formulations [12]. The results indicate that the use of two of the new models may lead to improvements to the bi-objective approach intro-

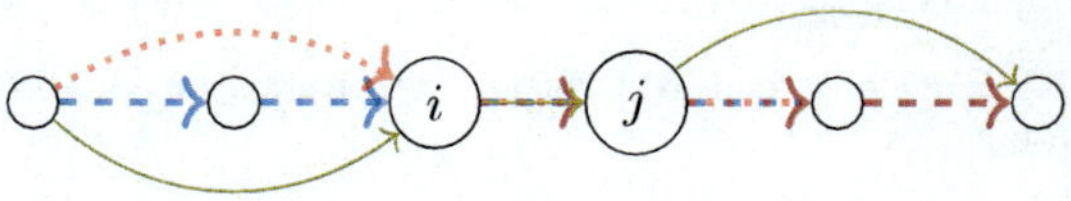

Fig. 1 Paths between a pair of nodes

duced in [14]. Thus, the present work addresses the K short and dissimilar problem incorporating the new models in the best frameworks identified in [14].

The rest of the text is organized as follows. In Sect. 2 the two ILP formulations are reviewed and the K short and dissimilar paths problem is defined. Section 3 is dedicated to extending the previous ILP formulations and to presenting an ε-constraint method to find non-dominated points for the resulting problems. Finally, Sect. 4 presents computational results and Sect. 5 concludes the text.

2 The K Short and Dissimilar Paths Problem

Let (N, A) be a directed graph with $|N| = n$ nodes and $|A| = m$ arcs, and let s and t denote given source and terminal graph nodes, respectively. The goal of the K dissimilar paths problem in (N, A) is to find a set of K paths from node s to node t, such that the paths in the set are "diverse" enough. This notion permits a wide range of interpretations and, thus, a number of dissimilarity measures have been proposed in the literature. Moghanni et al. [12] introduced three ILP formulations for the K dissimilar paths problem, which enhance previous models for the problem, proposed in [13]. In the following we extend our study to the two models in the first of these works that stood out, in terms of both times and dissimilarities, in order to test their behaviour in the bi-objective context. Next, we revisit these models, based on the strategy used:

- Minimizing the number of arc reuses in the K paths, called `MAR`. An arc is said to be reused when it appears in more than one path. Therefore, the number of arc reuses is the total number of times that an arc is used in more than one path.
- Minimizing the number of arc overlaps (counted pairwise) in the K paths, called `MAO`.

Figure 1 illustrates the difference between the two approaches. In the set of three paths depicted in the plot, arc (i, j) appears in all three of them, therefore it is reused twice. Moreover, there are three path overlaps, due to the overlap of the arc (i, j) for the paths with the solid and the dashed lines, the paths with the solid and the dotted lines, as well as the paths with the dashed and the dotted lines. The two formulations are shown next and are later extended to the bi-objective case.

Consider the usual single commodity flow variables, f_{ij}, accounting for the amount of flow (or the number of paths) traversing arc (i, j). Consider also variables u_{ij}, which correspond to the number of times that the arc $(i, j) \in A$ is reused in different paths, and variables w_{ij}, such that $w_{ij} = 1$ if and only if the arc (i, j) is used in at least one path, or $w_{ij} = 0$ otherwise, for any $(i, j) \in A$. Then the model `MAR`, to find K paths between nodes s and t with the minimum number of arc reuses, is as follows:

$$\min \; M_1 = \sum_{(i,j)\in A} u_{ij} \tag{1a}$$

$$\text{subject to} \quad \sum_{j\in N:(i,j)\in A} f_{ij} - \sum_{j\in N:(j,i)\in A} f_{ji} = \begin{cases} K & i = s \\ 0 & i \neq s,t \\ -K & i = t \end{cases} \tag{1b}$$

$$f_{ij} \leq K w_{ij}, \quad (i,j) \in A \tag{1c}$$

$$w_{ij} \leq f_{ij}, \quad (i,j) \in A \tag{1d}$$

$$u_{ij} = f_{ij} - w_{ij}, \quad (i,j) \in A \tag{1e}$$

$$f_{ij} \in \mathbb{N}_0, \; w_{ij} \in \{0,1\}, \; u_{ij} \geq 0, \quad (i,j) \in A \tag{1f}$$

The constraints (1b) are flow conservation constraints that define sets of K paths from node s to node t. The constraints (1c) and (1d) relate the w and the f variables, ensuring that, for any $(i, j) \in A$, $w_{ij} = 1$ if and only if (i, j) is used in some path, that is, if and only if $f_{ij} > 0$. In turn, constraints (1e) ensure that variables u_{ij} correspond to the number of times that arc $(i, j) \in A$ is reused in different paths. Therefore, the objective function counts the number of arc reuses in the K paths. This formulation is an aggregated version of the model used in [14]. Tests presented in [12] indicate that this model is faster than the one used [14], while maintaining the quality of the solutions regarding its dissimilarity.

The next formulation uses a set of discretized flow variables proposed in [7]. Consider the binary variables, g_{ij}^p, equal to 1 if arc (i, j) appears in exactly p paths, or to 0 otherwise, for any $(i, j) \in A$ and $p = 1, \ldots, K$. Then, the MAO formulation is the following:

$$\min \; M_2 = \sum_{(i,j)\in A} \sum_{p=1}^{K} \binom{p}{2} g_{ij}^p \tag{2a}$$

$$\text{subject to} \quad \sum_{j\in N:(i,j)\in A} \sum_{p=1}^{K} p\, g_{ij}^p - \sum_{j\in N:(j,i)\in A} \sum_{p=1}^{K} p\, g_{ji}^p = \begin{cases} K & i = s \\ 0 & i \neq s,t \\ -K & i = t \end{cases} \tag{2b}$$

$$\sum_{p=1}^{K} g_{ij}^p \leq 1, \quad (i,j) \in A \tag{2c}$$

$$g_{ij}^p \in \{0,1\}, \quad (i,j) \in A, \quad p = 1, \ldots, K \tag{2d}$$

The constraints (2b) define sets of K paths between the nodes s and t, and the constraints (2c) ensure that at most one of the g_{ij}^p variables associated to that arc is equal to 1, for any $(i, j) \in A$. The number of overlaps of an arc that appears in p paths corresponds to the number of times that it is shared by a pair of paths, for all possible pairs. This value is given by $\binom{p}{2}$, and therefore the objective function M_2 counts the number of arc overlaps for all the K paths (see [12]).

Formulation (2) is new to the bi-objective framework. However, results presented in [12] indicate that it outputs more dissimilar solutions, in spite of being slower than MAR in the single objective context.

Consider now that each arc in the network is associated with a cost $c_{ij} \in \mathbb{R}^+$, for any $(i, j) \in A$. The cost of a set of K paths defined as a solution of model MAR using the variables $f \in \mathbb{N}_0^m$ is given as the sum of their arc costs, that is

$$M_3^r = \sum_{(i,j)\in A} c_{ij} f_{ij}.$$

Similarly, given K solution vectors for model MAO, $g^p \in \{0, 1\}^m$, with $p = 1, \ldots, K$, the cost of the solution it represents is given by

$$M_3^o = \sum_{(i,j)\in A} c_{ij} \sum_{p=1}^{K} p\, g_{ij}^p.$$

The version of model (1) considering the simultaneous minimization of the paths cost and superposition can be formulated as:

$$\begin{aligned} \min \quad & M_3^r = \sum_{(i,j)\in A} c_{ij} f_{ij} \\ \min \quad & M_1 = \sum_{(i,j)\in A} u_{ij} \\ \text{subject to} \quad & (1b) - (1f) \end{aligned} \tag{3}$$

where the decision variables have the same meaning as before. Hereinafter this model will be designated as BMAR.

Considering now the decision variables used in formulation (2), the corresponding bi-objective version of that model, aiming at minimizing the cost and the number of arc overlaps, can be formulated as:

$$\begin{aligned} \min \quad & M_3^o = \sum_{(i,j)\in A} c_{ij} \sum_{p=1}^{K} p\, g_{ij}^p \\ \min \quad & M_2 = \sum_{(i,j)\in A} \sum_{p=1}^{K} \binom{p}{2} g_{ij}^p \\ \text{subject to} \quad & (2b) - (2d) \end{aligned} \tag{4}$$

This model will be designated as BMAO.

In general, the objective functions in formulation (3) and those in formulation (4) are conflicting. Therefore, rather than searching for optimal solutions, the goal of these problems is to search for efficient solutions, solutions for which there is no other that improves one of the objective functions without worsening the other. An ε-constraint method for finding solutions for these problems is described next.

3 Algorithm for Finding Non-dominated Sets of *K* Short and Dissimilar Paths

Considering two objective functions $f_1, f_2 : \mathbb{R}^n \to \mathbb{R}$, a bi-objective optimization problem can be defined as

$$\begin{array}{ll} \min & f(x) = (f_1(x), f_2(x)) \\ \text{subject to} & x \in X \end{array} \tag{5}$$

where $X \subseteq \mathbb{R}^n$ is a set of feasible solutions. In the approaches that we consider to the K short and dissimilar paths problem, BMAR and BMAO, the functions f_1 and f_2 correspond to a superposition function (either M_1 or M_2, respectively) and a cost function (M_3^r or M_3^o, respectively, both based on different expressions but with the same meaning).

A feasible solution of (5), $x^1 \in X$, is said to dominate another, $x^2 \in X$, if

1. $f_i(x^1) \leq f_i(x^2)$, for $i = 1, 2$, and
2. $f_i(x^1) < f_i(x^2)$, for at least one index $i \in \{1, 2\}$.

Additionally, a feasible solution $\bar{x} \in X$, is efficient, or Pareto optimal, if there is no other feasible solution, $x \in X$, which dominates $\bar{x}$. If the solution $\bar{x}$ is efficient, then its outcome vector $f(\bar{x})$ is called a non-dominated point. The set of all efficient solutions, denoted by X_E, is called the efficient set. The non-dominated set is the set of all non-dominated points, $Y_N = \{f(x) : x \in X_E\}$.

The ε-constraint method [8] is a well-known method able of finding the set of all non-dominated points for a bi-objective problem. The method consists of solving a sequence of single-objective problems, which optimize only one of the objective functions while satisfying a constraint where the remaining function is bounded by a variable parameter $\varepsilon > 0$.

The ε-constraint algorithm is particularly suited for the problem we are handling, given that the two objective functions are of different natures and this method does not imply an aggregation of the objectives. Without loss of generality, we consider that the objective function f_1 is minimized and the objective function f_2 is included in the constraints, which means that problem (5) is replaced by a sequence of ε-constraint problems

$$\begin{array}{ll} \text{minimize} & f_1(x) \\ \text{subject to} & x \in X \\ & f_2(x) \leq \varepsilon \end{array}, \quad \varepsilon \in \mathbb{R}. \tag{6}$$

Furthermore, updating ε as $f_2(\bar{x}) - \Delta$, with $\bar{x}$ a known feasible solution and $\Delta > 0$ a small number, guarantees an improvement of the second objective. With appropriate choices of ε all the non-dominated points of problem (5) can be found. The choice of the function to optimize and the one to include in the constraints, as well as of the strategy for updating the bound ε, may depend on the particular model.

The set of efficient solutions includes solutions that are minimal with respect to f_1 and with respect to f_2. However, there may be alternative optima for each of these objective functions which are weakly dominated by other solutions. In order to ensure that the first and the last solutions stored by Algorithm 1 are efficient, a lexicographic optimal solution for $\min_X(f_2, f_1)$ is computed. Such a solution can be found by minimizing function f_1, over the set of feasible solutions with a minimal value of f_2. A similar procedure can be followed for computing a lexicographic optimal solution for $\min_X(f_1, f_2)$. An outline of a generic ε-constraint method is given in Algorithm 1.

Algorithm 1: The ε-constraint method

1 $\bar{x} \leftarrow$ lexicographic optimal solution for $\min_X(f_2, f_1)$
2 $y_2^I \leftarrow f_2(\bar{x})$
3 $\bar{x} \leftarrow$ lexicographic optimal solution for $\min_X(f_1, f_2)$
4 $Y_E \leftarrow \{f(\bar{x})\}$
5 $\varepsilon \leftarrow f_2(\bar{x}) - \Delta$
6 **while** $\varepsilon \geq y_2^I$ **do**
7     $x^* \leftarrow$ optimal solution of problem (6)
8   **if** $f_1(x^*) > f_1(\bar{x})$ **then** $Y_E \leftarrow Y_E \cup \{f(\bar{x})\}$
9     $\bar{x} \leftarrow x^*$
10     $\varepsilon \leftarrow f_2(x^*) - \Delta$ // Update ε
11 $Y_E \leftarrow Y_E \cup \{f(\bar{x})\}$

The variable Y_E is a set that stores the non-dominated points of the problem as they are computed. For each value of ε, a new solution is found, x^*, and the parameter ε is updated according with its objective function value. The variable $\bar{x}$ is an auxiliary variable that stores the latest solution found until it is concluded whether it is efficient or it is just weakly efficient, and, therefore, dominated. The line 8 in the pseudo-code is a dominance test for solution $\bar{x}$. As a result, in case $\bar{x}$ is efficient, its image is included in the set Y_E.

The variable x^* is used to store potentially efficient solutions which are the optimal solutions of the ε-problems computed along Algorithm 1. The first point to be inserted in set Y_E is certainly non-dominated, because it corresponds to one of the lexicographic optimals for the objective functions. In the rest of the algorithm, the values of ε are strictly decreasing and, thus, so is the sequence $\{f_2(x^*)\}$. Therefore, if $\bar{x}$ temporarily stores an optimal solution of problem (6) and x^* is the next one, that is, obtained after updating ε, then $f_2(x^*) < f_2(\bar{x})$ and two situations may occur:

- $f_1(x^*) > f_1(\bar{x})$. This means that x^* does not dominate $\bar{x}$ and also that no forthcoming solution does. Therefore, $f\bar{x})$ is inserted in the set of efficient points, Y_E.
- $f_1(x^*) = f_1(\bar{x})$. This means that x^* dominates $\bar{x}$. Therefore, $\bar{x}$ is discarded and replaced by x^*.

Next we discuss the parameterization of Algorithm 1 when applied to models BMAR and BMAO. We first analyze the ranges of the objective functions M_1, M_3^r. In this case we have:

$$0 \leq M_1 \leq Kn,$$

because each path has at most $n-1$ arcs and each one can be reused up to $K-1$ times. Moreover,

$$1 \leq M_3^r \leq Kn \max_{(i,j)\in A} \{c_{ij}\},$$

because at most $K(n-1)$ arcs can be used in total. As a consequence, the range of M_1 is considerably tighter than the range of M_3^r, which results in fewer iterations of the `while` loop on line 6 of Algorithm 1 if we let f_2 be the first and f_1 be the second.

In the second model the cost objective is similar to the previous, thus $1 \leq M_3^o \leq Kn \max_{(i,j)\in A}\{c_{ij}\}$ holds, whereas, for the superposition objective,

$$0 \leq M_2 \leq \binom{K}{2} m = \frac{1}{2} K(K-1)m,$$

because each arc appears in at most K paths, thus producing up to $\binom{K}{2}$ overlaps. For this second problem the difference between the functions M_2 and M_3^o is not as sharp as for the former. Nevertheless, the upper bound on the number of arc overlaps is rarely met, and therefore in general the range of M_2 is still tighter than that of M_3^o leading to a similar conclusion as before.

If fixing M_3^r, M_3^o to optimize, the ε-constraint subproblems to solve are

$$\begin{aligned} \min \quad & M_3^r = \sum_{(i,j)\in A} c_{ij} f_{ij} \\ & \text{subject to (1b)} - \text{(1f)} \\ & \sum_{(i,j)\in A} u_{ij} \leq \varepsilon \end{aligned} \tag{7}$$

and

$$\begin{aligned} \min \quad & M_3^o = \sum_{(i,j)\in A} c_{ij} \sum_{p=1}^{K} p\, g_{ij}^p \\ & \text{subject to (2b)} - \text{(2d)} \\ & \sum_{(i,j)\in A} \sum_{p=1}^{K} \binom{p}{2} g_{ij}^p \leq \varepsilon \end{aligned} \tag{8}$$

with $\varepsilon > 0$. Then, the next result holds.

Proposition 1 *Any optimal solution of Problems (7) and (8) is formed by loopless paths.*

Problems (7) and (8) are close to an extension of K shortest path problems. Contrarily, the problems resulting from switching the roles of f_1 and f_2 are closer to the formulations (1) and (2) and may admit optimal solutions containing paths with loops. These loops can be discarded by applying an algorithm of $O(Km)$ time, without compromising the optimal value [12]. Nevertheless, it is simpler and more efficient to optimize the cost function than the superposition function. Additionally, both objective functions M_1, M_2 are intrinsically integer, whereas the same only happens for M_3^r, M_3^o if the arc costs are integer as well. Taking these considerations into account, the ε-algorithm for problems (3) and (4) is implemented by optimizing the cost function, M_3^r or M_3^o, respectively, and constraining the superposition function, M_1 or M_2, respectively. Moreover, the value $\Delta = 1$ is considered for updating the bound ε.

4 Empirical Study

In this section experiments to assess the empirical performance of the new formulations are presented. The purpose of these tests is trifold: i. to assess the efficiency of the ε-constraint algorithm; ii. to compare the performance of the formulations `BMAR` and `BMAO` for finding sets of K short and dissimilar paths; iii. to compare the results obtained by `BMAR` and `BMAO` with those obtained in [14].

The index used to assess dissimilarity, introduced in [6], measures the dissimilarity between paths p_i and p_j as:

$$D(p_i, p_j) = 1 - \frac{1}{2}\left(\frac{L(p_i \cap p_j)}{L(p_i)} + \frac{L(p_i \cap p_j)}{L(p_j)}\right) \tag{9}$$

where $L(p)$ denotes the number of arcs in sequence p. The dissimilarities vary from 0 to 1, the first for coincident paths and the latter for arc disjoint paths.

4.1 Tests Setup

Algorithm 1 was implemented for formulations `BMAR` and `BMAO`. The two models were coded in C, calling CPLEX 20.1 to solve the intermediate mixed-integer programs. The codes ran for two sets of instances, namely:

- Random graphs, $R_{n,m}$, with $n = 100, 500$ nodes, obtained generating randomly $m = dn$ arcs, with $d = 5, 10, 15$.
- Grid graphs, $G_{p,q}$, comprising the following sizes: $p \times q = 4 \times 36, 12 \times 12, 5 \times 45, 15 \times 15$. It is worth noting that in a grid topology there are always overlaps on sets of $K \geq 3$ paths. In addition, for these instances, the objective function M_2

Table 1 Description of the column headings

Heading	Description
$\bar{T}$	Average total run time, in seconds
$\lvert\bar{Y_E}\rvert$	Average number of computed non-dominated points
$\bar{N}$	Average number of solved subproblems
$\bar{f}^j_{\min}$ ($\bar{f}^j_{\max}$)	Average minimum (maximum) f_j in each set of paths, $j = 1, 2$
$\bar{D}_{\min}$ ($\bar{D}_{\max}$)	Average minimum (maximum) `AvDi` in each set of paths

measures paths dissimilarity exactly, given that all paths from the source to the destination nodes have the same length [12, 13].

In both cases each arc $(i, j) \in A$ was associated with a cost value, $c_{ij} \in \{1, 2, \ldots, 100\}$, uniformly obtained. The results in the following are averages obtained for finding sets of $K = 10, 20$ paths over 20 different instances generated for each dimension of these data sets. The tests ran on a 64-bit PC with an Intel® Core™ i9-10900 K Quad Core at 3.7 GHz with 128 GB of RAM. A time limit of 300 s was imposed for each subproblem solved along the generation of the non-dominated set. The test statistics are summarized in Table 1. Here `AvDi` is the average pairwise dissimilarity of each set of K paths.

4.2 Test Results

In the following we discuss the results of the application of Algorithm 1 to formulations (3) and (4) for the instances described above. We first consider the results for the formulation that minimizes the number of arc reuses.

The average results obtained by the code `BMAR` are shown on Tables 2 and 3. For the random instances, the number of non-dominated points seems to depend on n, while being quite resilient to the variation on m for the same value of n. For grids, the effect of the topology of the network overshadows the effect of the increase in size, as this number of non-dominated points is smaller in rectangles than in squares. This seems to be related with the number of arc reuses, and, thus, with the values $\bar{f}^2_{\min}$ and $\bar{f}^2_{\max}$. A path from 1 to n in a $p \times q$ grid contains $p + q - 2$ arcs and this number is minimized when $p = q$. For instance, such paths are shorter for 12×12 grids, with 22 arcs, and longer for 4×36 grids, with 38 arcs, even if both grids have the same number of nodes. This translates in a smaller number of arc reuses and, therefore, in higher average dissimilarity values, $\bar{D}_{\max}$, as will be seen later.

The number of iterations needed to find each non-dominated point measures the efficiency of the ε-constraint algorithm. The ratio $\bar{N}/\lvert\bar{Y_E}\rvert$ is either 1, if both values coincide, or greater otherwise, the latter situation corresponding to a less efficient behavior of the algorithm. In both tables, for random and grid networks, the parameter $\bar{N}/\lvert\bar{Y_E}\rvert$ is close to 1.

Table 2 Results for BMAR and $K = 10$

Instance	$\lvert\bar{Y}_E\rvert$	$\bar{N}$	$\bar{T}$	$\bar{N}/\lvert\bar{Y}_E\rvert$	$\bar{T}/\bar{N}$	$\bar{f}^1_{\min}$	$\bar{f}^1_{\max}$	$\bar{f}^2_{\min}$	$\bar{f}^2_{\max}$	$\bar{D}_{\min}$	$\bar{D}_{\max}$
$R_{100,500}$	24.68	25.68	2.20	1.040	0.08	906.8	1821.4	11.4	39.8	0.000	0.825
$R_{100,1000}$	28.50	29.55	3.99	1.034	0.13	555.5	1137.2	4.7	36.9	0.016	0.932
$R_{100,1500}$	27.25	28.30	3.71	1.036	0.13	387.0	888.3	2.7	31.9	0.010	0.971
$R_{500,2500}$	38.06	39.18	10.82	1.026	0.27	1239.4	2163.9	8.5	56.5	0.014	0.910
$R_{500,5000}$	33.90	34.90	38.83	1.364	1.11	761.0	1246.1	5.8	45.6	0.007	0.932
$R_{500,7500}$	34.65	35.80	62.93	1.029	1.75	481.0	911.1	2.6	42.4	0.016	0.977
$G_{12,12}$	139.20	140.25	13.33	1.007	0.09	6596.0	10363.8	40.0	196.9	0.010	0.920
$G_{4,36}$	102.15	103.45	11.74	1.010	0.11	14676.0	17337.5	214.0	342.0	0.000	0.675
$G_{15,15}$	187.75	189.05	38.03	1.011	0.20	7842.0	12629.9	40.0	251.3	0.005	0.937
$G_{5,45}$	166.75	168.35	42.66	1.006	0.25	18660.5	22169.4	228.0	431.8	0.001	0.758

Table 3 Results for BMAR and $K = 20$

Instance	$\lvert \bar{Y_E} \rvert$	$\bar{N}$	$\bar{T}$	$\bar{N}/\lvert \bar{Y_E} \rvert$	$\bar{T}/\bar{N}$	$\bar{f}^1_{\min}$	$\bar{f}^1_{\max}$	$\bar{f}^2_{\min}$	$\bar{f}^2_{\max}$	$\bar{D}_{\min}$	$\bar{D}_{\max}$
$R_{100,500}$	40.84	41.84	3.20	1.024	0.08	1 813.7	3 393.9	37.37	84.00	0.000	0.678
$R_{100,1000}$	53.20	54.25	6.71	1.020	0.12	1 111.0	2 518.2	19.00	78.40	0.008	0.874
$R_{100,1500}$	51.90	52.95	6.53	1.020	0.12	774.0	2 085.3	13.05	67.35	0.005	0.925
$R_{500,2500}$	70.06	71.25	17.91	1.017	0.25	2 478.8	4 396.2	34.00	119.62	0.007	0.864
$R_{500,5000}$	70.95	71.95	77.99	1.014	1.08	1 522.0	2 962.0	18.20	96.60	0.003	0.911
$R_{500,7500}$	71.35	72.50	153.46	1.016	2.12	962.0	2 139.8	12.55	89.85	0.008	0.957
$G_{12,12}$	219.05	220.35	16.81	1.006	0.08	13 192.0	22 223.4	180.0	416.9	0.005	0.856
$G_{4,36}$	131.80	133.00	11.08	1.009	0.08	29 352.0	35 355.2	564.0	722.0	0.000	0.588
$G_{15,15}$	327.55	329.10	58.59	1.005	0.18	15 684.0	28 296.0	180.0	531.3	0.003	0.886
$G_{5,45}$	205.60	207.25	44.75	1.008	0.22	37 321.0	44 534.2	668.0	911.8	0.000	0.645

The average time for solving each subproblem increases with the size, both in random and grid networks. The results are different in square and in rectangular grids with the same number of nodes, the latter being harder to solve. This fact, together with the increase in the number of solutions that need to be found, results also in an increase of the total run time.

As expected, the range of the cost, f_1, is larger than the range of the number of arc reuses, f_2. Also, the values of both objective functions increase with n, and decrease as m increases in networks with the same number of nodes. Results for rectangular grids, show higher cost and arc reuse values, when compared to the square instances, and this effect overcomes the effect of the size of the network. The average minimum dissimilarity was always nearly 0. This is explained by the fact that most of the determined sets contains at least one solution formed by paths that coincide with the shortest. The average maximum dissimilarity, on the other hand, lies between 0.825 and 0.977 (0.678 and 0.957) in random networks and between 0.675 and 0.937 (0.588 and 0.886) in grid networks, when $K = 10$ ($K = 20$). As a general trend, and as expected, the dissimilarities increase with m and are bigger in the square networks.

Tables 4 and 5 summarize the results of the code BMAO. First, it should be pointed out that not all the subproblems associated to the grid instances were solved to optimality within the imposed time limit. For random networks, the average number of non-dominated points computed when using BMAO depended mainly on n. For grids, using BMAO led to a higher number of non-dominated points for the rectangular grids. As before, the ratio $\bar{N}/|\bar{Y}_E|$ is very close to 1, confirming the efficiency of the proposed method.

For the random instances, the run times of the subproblems associated to using Algorithm 1 with BMAO increase with size. These values are also greatly affected by the number of paths to find, K. On the other hand, for the grid instances, the values of $\bar{T}/\bar{N}$ indicate that the problem is significantly harder to solve on rectangular grids than on square grids.

The range of the average number of arc overlaps of the non-dominated solutions obtained by using BMAO in the above algorithm is tighter than the range of the average cost of the same solutions. Also as expected, the values of both $\bar{f}^2_{\min}$ and $\bar{f}^2_{\max}$ increase with n and decrease with m for the random instances, and are significantly higher for rectangular grids than for square grids. The average dissimilarities registered for BMAO vary as predicted: for random networks the average dissimilarities decrease with n, while they increase with m for the same n; the average dissimilarities are bigger for the square grids than for the rectangular grids. Also as predicted, the average maximum dissimilarity decreases with K for all the instances. Again, $\bar{D}_{\min}$ is nearly zero. In addition, the values of $\bar{D}_{\max}$ lie between 0.869 and 0.985 (0.847 and 0.975) in random networks, and between 0.815 and 0.943 (0.782 and 0.915) in grid networks, when $K = 10$ ($K = 20$).

The differences between BMAR and BMAO are summarized in Fig. 2. When $K = 10$ the latter model produces around the double of the solutions of BMAR (more in grid instances), with a cost range that is 1.05 times wider than that obtained by BMAR, corresponding to an improvement of the maximum dissimilarity of the solutions of

Table 4 Results for BMAO and $K = 10$

Instance	$\lvert\bar{Y}_E\rvert$	$\bar{N}$	$\bar{T}$	$\bar{N}/\lvert\bar{Y}_E\rvert$	$\bar{T}/\bar{N}$	$\bar{f}^1_{\min}$	$\bar{f}^1_{\max}$	$\bar{f}^2_{\min}$	$\bar{f}^2_{\max}$	$\bar{D}_{\min}$	$\bar{D}_{\max}$
$R_{100,500}$	60.05	61.52	22.89	1.033	0.37	906.8	1975.2	27.5	198.9	0.000	0.869
$R_{100,1000}$	68.90	71.45	38.60	1.029	0.54	555.5	1202.8	7.0	177.3	0.040	0.961
$R_{100,1500}$	66.45	70.05	45.37	1.061	0.64	387.0	935.9	2.9	158.3	0.023	0.983
$R_{500,2500}$	80.56	83.93	225.37	1.031	2.68	1239.4	2160.8	20.2	277.3	0.036	0.926
$R_{500,5000}$	70.10	72.95	587.17	1.043	8.04	761.0	1325.3	12.1	222.5	0.019	0.950
$R_{500,7500}$	67.55	70.85	856.62	1.044	12.09	481.0	941.2	3.5	206.7	0.030	0.985
$G^*_{12,12}$	372.90	387.65	108.72	1.040	0.28	6596.0	10250.9	72.0	961.3	0.029	0.927
$G^*_{4,36}$	522.85	547.50	309.82	1.050	0.56	14676.0	18546.2	316.0	1710.0	0.000	0.815
$G^*_{15,15}$	519.50	542.25	249.44	1.042	0.46	7842.0	12459.6	72.0	1242.3	0.014	0.943
$G^*_{5,45}$	733.65	762.15	596.85	1.038	0.78	18660.5	24496.5	266.0	2155.0	0.002	0.877

*: subproblems interrupted after 300 s

Table 5 Results for BMAO and $K = 20$

Instance	$\lvert\bar{Y}_E\rvert$	$\bar{N}$	$\bar{T}$	$\bar{N}/\lvert\bar{Y}_E\rvert$	$\bar{T}/\bar{N}$	$\bar{f}^1_{\min}$	$\bar{f}^1_{\max}$	$\bar{f}^2_{\min}$	$\bar{f}^2_{\max}$	$\bar{D}_{\min}$	$\bar{D}_{\max}$
$R_{100,500}$	215.84	222.84	215.18	1.032	0.97	1 813.7	4 448.6	137.4	840.0	0.000	0.847
$R_{100,1000}$	248.40	262.50	369.52	1.057	1.41	1 111.0	2 847.4	41.4	751.3	0.037	0.942
$R_{100,1500}$	244.95	258.65	599.30	1.056	2.32	774.0	2 358.8	20.6	669.0	0.020	0.969
$R_{500,2500}$	303.50	315.62	4 664.03	1.040	14.78	2 478.8	4 894.3	104.9	1 172.8	0.035	0.910
$R_{500,5000}$	263.80	280.90	8 803.35	1.066	31.34	1 522.0	3 176.8	61.1	941.1	0.018	0.938
$R_{500,7500}$	158.20	175.30	4 426.07	1.108	25.25	962.0	18 400.2	25.9	874.9	0.029	0.975
$G^*_{12,12}$	1 334.20	1 431.60	785.90	1.073	0.55	13 192.0	21 560.5	448.0	4 065.0	0.028	0.893
$G^*_{4,36}$	1 716.65	1 824.50	1 711.13	1.063	0.94	29 352.0	38 259.5	1 572.0	7 220.0	0.000	0.782
$G^*_{15,15}$	1 873.20	2 007.10	1 737.60	1.071	0.87	15 684.0	27 350.6	452.0	5 249.2	0.013	0.915
$G^*_{5,45}$	2 197.40	2 334.95	3 570.59	1.063	1.53	37 321.0	48 945.1	1 530.0	9 100.0	0.002	0.832

*: subproblems interrupted after 300 s

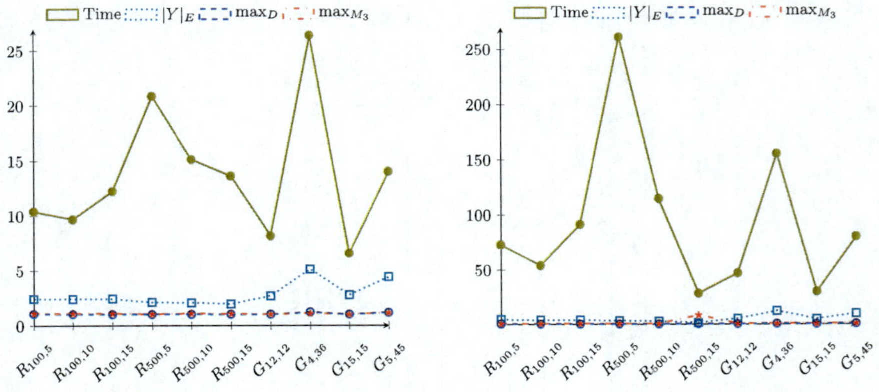

(a) $K = 10$ (b) $K = 20$

Fig. 2 Comparison between the results of BMAO and BMAR (BMAO/BMAR)

1.3 times that obtained by BMAO. In terms of the run time, BMAO is always slower than BMAR: from 10 to 21 times slower for random instances and from 7 to 26 times slower for grids. The same general trend is observed for $K = 20$, with steeper differences in the run times of the two codes.

The application of the codes BMAR and BMAO results in two lists of sets of K paths, which are efficient in terms of the cost and of the number of arc reuses/overlaps, respectively. These lists are associated with approximations to the Pareto front for the cost and the dissimilarity of the K paths, therefore in the following we assess their quality. Let $P_a = \{(f_1^1, D^1), (f_1^2, D^2), \ldots, (f_1^{m_a}, D^{m_a})\}$ be a list of cost and dissimilarity pairs obtained by the previous approaches. We consider different aspects for assessing the quality of this set. Namely:

- Average distance between the set P_a and a point, (z_c^*, z_d^*):

$$\bar{d}_a = \sum_{i=1}^{m_a} \frac{\|(f_1^i, D^i) - (z_c^*, z_d^*)\|}{m_a},$$

  where $\|.\|$ stands for the Euclidean norm. The ideal point is chosen as the reference point. Two variants of this metric are used: total distance, considering all the points in P_a, $\bar{d}_a^T$, and partial distance, restricting the points generated by the approach to the intersection of the regions covered by both approaches, $\bar{d}_a^P$.
- Purity of the set P_a [2], defined by $r_a = |P_a \cap P|/|P_a|$, where P is a set of non-dominated points. This metric aims at evaluating the extent to which the set of points obtained with one of the models is dominated by the other with respect to cost and dissimilarity. We consider that P is given by all the points obtained by BMAR and BMAO that are not dominated amongst them.

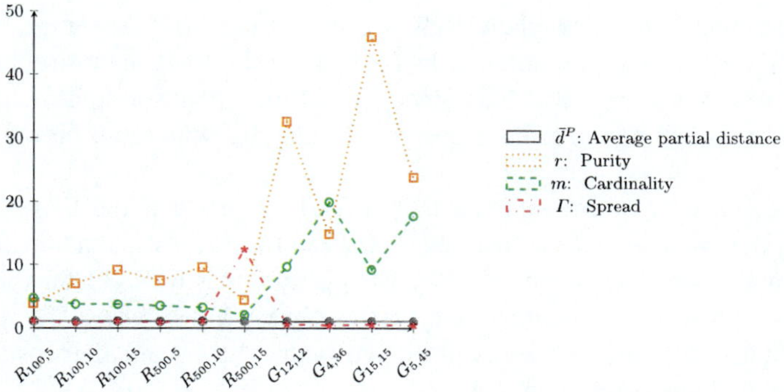

Fig. 3 Comparison of the sets of non/dominated cost-dissimilarity pairs obtained by BMAO and BMAR (BMAO/BMAR)

- Spread of the set P_a [4], given by $\Gamma_a = \max_{i=1,\dots,m_a} \left\{\max\{f_1^{i+1} - f_1^i, D_{i+1} - D_i\}\right\}$, measuring how well the non-dominated points are distributed, namely, measuring the maximum distance between consecutive points. Cost and dissimilarity are normalized when calculating this parameter, in order to consider values with similar ranges.

Smaller average distances correspond to sets of solutions that are closer to the reference point, and smaller spreads correspond to sets of points that are closer from one another. Greater purity values correspond to sets of solutions with more points that are not dominated by any of the those that are known. A comparison of these values is shown in Fig. 3 and a more detailed summary is shown in Table 6.

The number of cost/dissimilarity pairs that are not dominated is considerably smaller than the number of solutions output, from 60 to 77% with BMAR, and around

Table 6 Comparison of the fronts for $K = 20$

	BMAR					BMAO				
Instance	d^T_{BMAR}	d^P_{BMAR}	r_{BMAR} (%)	m_{BMAR}	Γ_{BMAR}	d^T_{BMAO}	d^P_{BMAO}	r_{BMAO} (%)	m_{BMAO}	Γ_{BMAO}
$R_{100,500}$	490.0	31.9	26.09	25.0	155.9	801.3	34.4	99.88	117.5	173.2
$R_{100,1000}$	465.8	14.9	14.42	35.3	136.9	534.4	15.9	99.93	131.6	102.4
$R_{100,1500}$	419.6	10.4	11.01	39.7	86.5	435.0	11.6	99.96	147.3	89.2
$R_{500,2500}$	628.4	17.5	13.51	46.0	136.2	689.6	18.3	100.00	160.0	112.4
$R_{500,5000}$	456.0	10.0	10.52	47.9	96.8	449.6	10.6	99.77	153.6	101.6
$R_{500,7500}$	402.6	8.2	23.13	47.2	80.4	396.4	9.0	99.81	96.9	991.7
$G^*_{12,12}$	2 137.2	8.6	3.08	140.0	274.9	2 014.3	8.7	100.00	1 352.1	132.4
$G^*_{4,36}$	1 461.3	15.4	6.77	84.8	325.6	3 117.3	16.0	100.00	1 690.6	97.6
$G^*_{15,15}$	3 053.0	9.0	2.18	198.2	325.8	2 584.8	9.5	100.00	1 824.8	136.8
$G^*_{5,45}$	1 868.3	14.6	4.21	123.4	355.3	3 878.0	15.2	100.00	2 182.7	123.0

50% on random instances and around 90% in grids with BMAO. Like for the output of Algorithm 1, the lists of points obtained by BMAR were shorter than those obtained by BMAO. In random networks BMAO obtained 2 to 5 times more solutions than BMAR. The difference was bigger in grids, especially rectangular, with ratios between 9 and 21.

The percentage of non-dominated pairs found by BMAR if the lists of points obtained by the two approaches is merged is between 13 and 29% in random instances and even smaller on grids, from 2 to 7%. The purity rate is much better for BMAO, even if it provided more solutions, with values always above 98%. The reason is that BMAR finds solutions with worse dissimilarity than BMAO and their costs are not smaller than with the other approach.

The average distance to the ideal point varies and is sometimes bigger for BMAR, while for others the opposite happens. These results are not fully conclusive because the lists produced by both codes have different ranges. The partial distance constrains the points to those with common range. In this case the values are always 3 to 11% bigger for BMAO than for BMAR, except for 12×12 grids. These results are due to the fact that more points are considered when using BMAO, even when restricting the range. Finally, the spread/distance between consecutive points is consistently smaller for BMAO than for BMAR, except for random instances with $n = 50$ and $m = 7500$, therefore we can conclude that those frontiers of points are more complete and uniform than the latter.

5 Final Remarks

This work addressed the problem of finding sets of K paths that are as dissimilar as possible, while also minimizing the total cost. A bi-objective approach for finding non-dominated points to this problem was described. The approach consists of an ε-constraint method, which is parameterized according to the problem, and then empirically tested for random and grid instances artificially generated. The computational results were discussed.

The code BMAR, based on minimizing the number of arc reuses, was faster than BMAO, based on minimizing the number of arc overlaps, because it required computing fewer non-dominated points and it solved easier subproblems. However, code BMAO produced more solutions, with better dissimilarities, and often dominated those output by BMAR in terms of their cost and dissimilarity.

Finally, comparing the new results to those presented in [14], it can be seen that BMAR speeds up its disaggregated version by an average factor of 22% for random networks and 32% for grids (bigger for networks with fewer nodes). On the other hand, BMAO produces solutions with higher average dissimilarities than the best model introduced in [14], although being slower to solve the same instances.

Acknowledgements The work of MP and AM was partially financially supported by the Portuguese Foundation for Science and Technology (FCT) under project grants UID/MAT/00324/2020 and UID/MULTI/00308/2020.

6 Test Results

See Tables 2, 3, 4, 5 and 6.

References

1. Akgün, V., Erkut, E., Batta, R.: On finding dissimilar paths. Eur. J. Oper. Res. **121**, 232–246 (2000)
2. Bandyopadhyay, S., Pal, S., Aruna, B.: Multiobjective gas, quantitative indices, and pattern classification. IEEE Trans. Syst. Man Cybern. Part B: Cybern. **34**(5), 2088–2099 (2004)
3. Carotenuto, P., Giordani, S., Ricciardelli, S.: Finding minimum and equitable risk routes for hazmat shipments. Comput. Oper. Res. **34**(5), 1304–1327 (2007)
4. Deb, K., Pratap, A., Agarwal, S., Meyarivan, T.: A fast and elitist multiobjective genetic algorithm: NSGA-II. IEEE Trans. Evol. Comput. **6**, 182–197 (2002)
5. Dell'Olmo, P., Gentili, M., Scozzari, A.: On finding dissimilar Pareto-optimal paths. Eur. J. Oper. Res. **162**, 70–82 (2005)
6. Erkut, E., Verter, V.: Modeling of transport risk for hazardous materials. Oper. Res. **46**, 625–642 (1998)
7. Gouveia, L.: A 2n constraint formulation for the capacitated minimal spanning tree problem. Oper. Res. **43**(1), 130–141 (1995)
8. Haimes, Y., Lasdon, L., Wismer, D.: On a bicriterion formulation of the problems of integrated system identification and system optimization. IEEE Trans. Syst. Man Cybernet. SMC-1:296–297 (1971)
9. Johnson, P.E., Joy, D.S., Clarke, D.: Highway 3.01, an enhancement routing model: program, description, methodology and revised user's manual. Working Paper, Oak Ridge National Laboratories, Washington, DC (1992)
10. Kuby, M., Zhongyi, X., Xiaodong, X.: A minimax method for finding the *k* best "differentiated" paths. Geogr. Anal. **29**, 298–313 (1997)
11. Martí, R., González-Velarde, J., Duarte, A.: Heuristics for the bi-objective path dissimilarity problem. Comput. Oper. Res. **36**, 2905–2912 (2009)
12. Moghanni, A., Pascoal, M., Godinho, M.T.: Finding *K* dissimilar paths: single-commodity and discretized flow formulations. Comput. Oper. Res. **147**, 105939 (2022)
13. Moghanni, A., Pascoal, M., Godinho, M.T.: Finding dissimilar paths using integer linear formulations. Technical report 21–33, Department of Mathematics, University of Coimbra (2021)
14. Moghanni, A., Pascoal, M., Godinho, M.T.: Finding *K* shortest and dissimilar paths. Int. Trans. Oper. Res. **29**, 1573–1601 (2021)
15. Thyagarajan, K., Batta, R., Karwan, M., Szczerba, R.: Planning dissimilar paths for military units. Mil. Oper. Res. **10**, 25–42 (2005)

Time Windows Vehicle Routing Problem to On-Time Transportation of Biological Products on Healthcare Centres

Maria Teresa Pereira, Marisa Oliveira, Fernanda Amélia Ferreira, Alcinda Barreiras, and Liliana Carneiro

Abstract This paper addresses a Vehicle Routing Problem (VRP) applied to the field of healthcare. Biological products are collected from patients at the local healthcare centers and transported to hospital laboratories for further processing and analysis. This paper analyses and determines a set of vehicle routes to perform on-time transportation of biological products from local healthcare centers to the main hospital, considering all technical issues. We sought to develop a solution to the Vehicle Routing Problem with Pickups and Deliveries (VRPPD) to effectively collect biological products, and parallelly deliver medical supplies to local healthcare units—gloves, masks, sanitation accessories, and disposable tools. We also aimed to implement a solution suitable for a larger cluster of healthcare centers. The mathematical model allowed for an efficient route design, considering distances, service times, travel times, total route time, and vehicle availability for other tasks. The mathematical model (VRPPDW) presents a feasible improvement to the solution currently used by the healthcare units. It allows for pickup and delivery of other items as required, and can be adapted if other collection points are to be added, providing a strong route and service times optimization. We were able to achieve a 95-min reduction, thus saving

M. T. Pereira (✉) · M. Oliveira · A. Barreiras · L. Carneiro
School of Engineering of Porto (ISEP), Polytechnic of Porto, Porto, Portugal
e-mail: mtp@isep.ipp.pt

M. Oliveira
e-mail: mjo@isep.ipp.pt

A. Barreiras
e-mail: asb@isep.ipp.pt

L. Carneiro
e-mail: 1161295@isep.ipp.pt

M. T. Pereira
INEGI - Institute of Mechanical Engineering and Industrial Management, Porto, Portugal

F. A. Ferreira
UNIAG, School of Hospitality and Tourism of Polytechnic Institute of Porto, Porto, Portugal
e-mail: faf@esht.ipp.pt

J. P. Almeida et al. (eds.), *Operational Research*, Springer Proceedings in Mathematics & Statistics 411,
https://doi.org/10.1007/978-3-031-20788-4_11

€2,222.64 per year. This solution required no further investment thus avoiding any reallocation available resources.

Keywords Vehicle routing problem · Time windows · Biological products transportation · Healthcare centres · Decision support

1 Introduction

In this work a Vehicle Routing Problem (VRP) is presented to design an optimal route for a vehicle fleet to perform requests at minimum cost, [1]. There are several papers demonstrating the employment of VRP to healthcare services, e.g. patient transportation for surgery and other medical procedures, applied to short lifespan items such as blood/plasma, or medication. These have critical delivery time windows, and hence require maximum route optimization. In the case of a route made for the collection of blood, this means covering all possible health units, taking into consideration the time-limit for keeping blood without centrifugation, ensuring the shortest possible time with the available resources. This study introduces a solution of the Vehicle Routing Problem with Pickups and Time Windows (VRPPTW) for a Portuguese Local Health Unit (LHU). A set of vehicle routes to perform on-time transportation of biological products from the healthcare centers had to be determined—from the collection points to the hospital laboratory for further processing and analysis. Thus, we sought to improve the existing solution to the Vehicle Routing Problem with Pickups and Deliveries (VRPPD) to effectively collect biological products, and parallelly deliver medical supplies to local healthcare centers—gloves, masks, sanitation accessories, and disposable tools. We also aimed to implement a solution suitable for larger clusters of healthcare centers. Expectations were to validate the model (VRPPTW) developed for the LHU and to generate optimized routes to allow both transportation of biological products to the laboratory, within specified time windows, and supply local healthcare centers. This entailed increasing the complexity of the existing management model by considering vehicle capacities, management of same time pickups and deliveries as well as enlarged network of LHU.

2 Literature Review

The VRP purpose is to define a set of vehicles, the routes and the clients to visit in well-defined locations, [2]. It is one of the most studied problems in the literature because of its wide applicability and importance in determining efficient strategies to reduce operational costs in distribution networks, [3]. This problem is comprised of several variants, namely the transportation of goods, which can contemplate the situation of delivery and collection of goods, Pickup and Delivery Problems (PDP), that is, objects have to be transported between an origin and a destination. These can be classified into three groups according to Oliveira et al. [4]:

- n origins to m destinations, where any vertex can serve as a source or as a destination for any commodity;
- 1 origin to n destinations and 1 final destination;
- 1 origin to 1 destination.

Since the original introduction of the concept, researchers have experimented with many models and solution methods which made VRP evolve to more complex variants, with a greater number of constraints.

- VRPPD—Vehicle Routing Problem with Pickups and Deliveries.
  The idea is to find a set of optimal routes, for a fleet of vehicles, to serve a set of transportation requests. Each vehicle from the fleet has a given capacity, a start location, and an end location. Each transportation request is specified by a load to be transported, an origin, and a destination. In other words, the problem deals with the design of optimal routes in order to visit all pickup and delivery locations, and satisfy precedence (each pickup location has to be visited prior to visiting the corresponding delivery location) and pairing constraints (one vehicle has to do both the pickup and the delivery of the load of one transportation request) [5].
- VRPTW—Vehicle Routing Problem with Time Windows.
  With this type of problem, capacity constraint remains an issue to be considered, and each customer is associated with a time interval $[a_i, b_i]$, called the time window, and with a time duration, s_i, the service time. These constraints restrict the times at which a customer is available to receive a delivery. This problem is often common in real-world applications, since the assumption of complete availability over time of the customers made in CVRP is often unrealistic, [6, 7].
- VRPSD—Vehicle Routing Problem with Split Deliveries.
  This type of problem suggests that the customer can be served by more than one vehicle. In this way, it is possible to control deliveries from customers that exceed the capacity of the distribution vehicle, [8].
  Considering the main characteristics of the types of problems described, this project work retained mainly VRPPTW and VRPPD.
- CVRP—The Capacitated Vehicle Routing Problem.
  This problem is a Vehicle Routing Problem (VRP) as it includes capacity constraints—the limited space of each vehicle and a depot. The objective of solving this problem is to find the shortest route, while every customer with a certain demand must be visited exactly once, [9].

Four main components constitute a Vehicle Routing Problem:

- the network which is generally described by the graph;
- the sites to be visited represented as customers which have a specific request often called a demand;
- the fleet of vehicles represented by the mobiles performing a task;
- the depot(s) usually from where the vehicles start and come back, [10].

Mathematically there is a directed graph $G = (V, A)$ where $V = \{0, 1, ..., n\}$ is a set of $n + 1$ nodes and A is a set of arcs. Node 0 represents a depot while the remaining node set corresponds to customers. To each arc $(i, j) \in A$ is assigned a distance or cost c_{ij}. Binary decision variable x_{ij} is set to one if and only arc (i, j) is used in the solution. The problem can be formulated as: the aim of basic version also called capacitated VRP (CVRP) is to determine the optimal set of routes to be performed by a fleet of capacitated vehicles to serve the demand of a given customer set [10]. A set of homogeneous vehicles each with a capacity of Q, located at a central depot and a set of customers with known locations and demands to be satisfied by deliveries from the central depot, each vehicle route must start and end at the central depot, and the total customer demand satisfied by deliveries on each route must not exceed the vehicle capacity, Q. The total cost is usually proportional to the total distance travelled if the number of vehicles is fixed and may also include an additional term proportional to the number of vehicles used if the number of routes may vary, [6].

The routing of blood collection vehicles has been studied by Özener [11], that applied a VRP model for improving the platelet supply in the blood supply chain. They analyze the routing decisions in such a setting and propose an integrated clustering and routing framework to collect and process the maximum number of donations for platelet production. Other related works using VRP concern bio-medical waste collection and home healthcare. Faizal [12] applied a VRP using Particle Swarm Optimization Algorithm (PSO) to optimize collection times. Cissé [13] presented a review of the relevant routing scheduling problems related with home healthcare.

3 Problem Description

A Local Health Unit (LHU) in greater Oporto is comprised of a central hospital and seven associated local healthcare centers (LHC), referred to here by HA, HB,..., HG. The LHU applies a certain procedure to conduct clinical analysis. The collection of samples is carried out in healthcare centers, and after such biological products are sent to the Clinical Pathology services in the Hospital (H), which has a central laboratory for their examination, but only works in the morning. The idea is to distribute patients for clinical analysis by geographic area, instead of concentrating them at a single point. Local residents can currently do their clinical analysis at the geographically assigned LHC. 70% of collected samples correspond to blood which need to strictly comply to government regulation (internal ordinance, no. 8, Official Gazette 166/2014 of August 21st 2014). Hence, after collecting blood and before transporting it to the laboratory, the sample elements (red blood cells, platelets, plasma) have to be separated - the centrifugation process. At the time, this process was only available at the hospital. The blood collected in the collection points (LHC) which do not have centrifuges had to be processed within two hours after collection, with a 30-minute tolerance. The use of centrifuges allows the samples to keep a sufficient quality for clinical analysis up to 4 h after undergoing the centrifugation process. The problem is that the initially established route for transporting non-

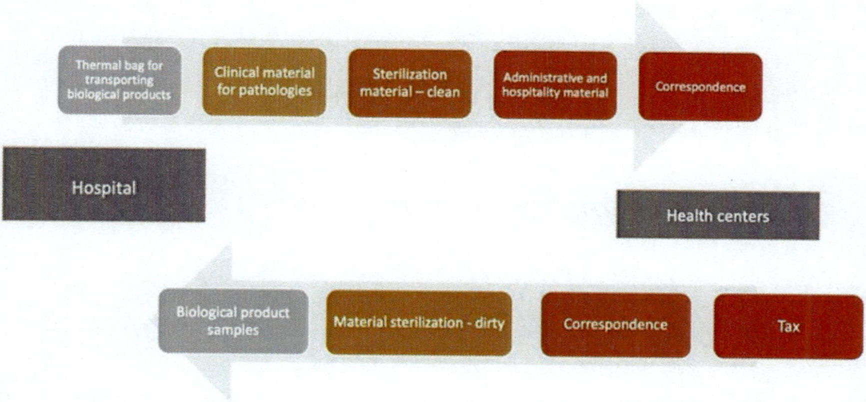

Fig. 1 Material flow of the LHU

centrifugated samples (from HD-HG) does not satisfy the required delivery time limit of 2.5 h. Considering LHU network, transportation restrictions and blood collection time frames of the LHC , we intended to develop a mathematical model to generate an optimal route for transporting biological products within specified time limits. In the route, along the biological products pickup, pickup and delivery of all kinds of programmed or urgently needed materials, such as sterilization material, cleaning material, disposables, and son on, is also required Fig. 1.

4 The Mathematical Model

The model is based on the vehicle routing with time windows formulation for the VRPTW, [14] which is based on a directed graph $G = (NHU \cup \{0, nHU + 1\}, A)$, where the depot is represented by the two nodes 0 and $nHU + 1$. The routes start from node 0 and end at node $nHU + 1$. In these nodes the service time is zero, this means that $ts_0 = ts_{nHU+1} = 0$. $NHU = \{1, 2, ..., nHU\}$ is the set of LHC, and $A = \{(i, j) : i \in NHU \cup \{0\}, j \in NHU \cup \{nHU + 1\}, i \neq j\}$ are the arcs.

The parameters are:

NHU Health Units $\{1, 2, \ldots, nHU\}$,
a_i lower limit of time window of $i, i \in NHU$,
b_i lower upper of time window of $i, i \in NHU$,
tt_{ij} travel time from i to $j, i \in NHU \cup \{0\}, j \in NHU \cup \{nHU + 1\}$
ts_i service time, corresponds to the time the driver spends collecting samples from each $i, i \in NHU$.

The decision variables are:

$$x_{ij} = \begin{cases} 1 \text{ } \textit{if the vehicle travels from i to j} \\ 0 \text{ } \textit{otherwise} \end{cases}$$

$$i \in NHU \cup \{0\},\ j \in NHU \cup \{nHU + 1\}$$

$$s_i = \textit{instant when the vehicle starts service at i}\ ,\ i \in NHU \cup \{nHU + 1\}$$

The objective function (1) aims to minimize the completion time of the tour.

$$\min\ s_{nHU+1} \tag{1}$$

subject to:

$$s_i - tt_{0i} \cdot x_{0i} \geq 0 \qquad \forall i \in NHU \tag{2}$$

$$s_i \geq a_i \qquad \forall i \in NHU \tag{3}$$

$$s_i + ts_i - s_j + (b_i - a_j + tt_{ij}) \cdot x_{ij} \leq b_i - a_j \ \ \forall i \in NHU,\ j \in \{NHU, i \neq j\} \tag{4}$$

$$\sum_{i \neq j, i \in NHU \cup \{0\}} x_{ij} = 1 \qquad \forall j \in NHU \tag{5}$$

$$\sum_{j \neq i, j \in NHU \cup \{nHU+1\}} x_{ij} = 1 \qquad \forall i \in NHU \tag{6}$$

$$s_i + ts_i \leq b_i \qquad \forall i \in NHU \tag{7}$$

$$s_i + ts_i + tt_{i0} \leq s_{nHU+1} \qquad \forall i \in NHU \tag{8}$$

The first constraint (2) initiates from the arrival time of the vehicle at the first LHC. Constraints (3) and (7) define a time window for each LHC within which the vehicle may arrive and collect samples. The constraint (4), when the vehicle travels between two LHC i and j, becomes $ts_i + tt_{ij} \leq s_j - s_i$ which ensures the difference between the arrival times at i and j is at least the travel time between these locations and the service time at i. The constraint (4) is also used to avoid sub tours. Constraints (5) and (6) ensure that the vehicle visits each LHC only once. The tour duration is determined by the constraint (8).

5 Results

The model was implemented with software IBM ILOG CPLEX Optimization Studio 21.1.0.0. After formulating the problem, and based on the existing routes, we realized that two loops are needed to meet the collection requirement of biological products. The routes were only defined for a single vehicle.

The LHC in HG, HD, HE and HF do not have a centrifuge. Table 1 provides information regarding health units' timeframes for biological products harvest.

As these clinics do not have centrifuges installed, there is a time limit of 2 h (+30 min tolerance) for samples delivery from the moment of blood harvest to the laboratory. Due to this constraint, collecting all the samples in one cycle results in having a part of them unusable for analysis. For the sake of efficiency, and to ensure that all samples were usable, we decided to introduce to 2 loops per route. The LHC which had the earliest time windows were included in the 1st loop to ensure the 2-h time limit (+ 30 min tolerance). The remainder LHC, with later blood collection timeframes, i.e., 10 to 11 pm, were in included in the 2nd loop, to ensure the legal time limit is respected and samples are delivered to the hospital before the closing of the laboratory which, as previously referred, only operates in the morning. It was noted that since the first LHC unit opens at 07:30, the time limit for arrival at the laboratory is, at the latest, at 10:00. However, to free the vehicle for the second loop, it should arrive as close as possible to 10:00 am. For the first simulation, we decided to use mid time in the harvesting period as the lower limit of the time window a_i and closing time of the units as the upper limit b_i. Table 2 represents the boundaries for the first loop, scenario 1. Note that LHU is the depot.

Table 1 Harvesting period in Health Units w/o centrifuges

LHC	Opening hour (h)	Closing hour (h)
HG	08:00	11:00
HD	07:30	11:00
HE	08:00	11:00
HF	08:00	10:30

Table 2 Boundaries for time windows

i	Health unit	a_i (hh:mm)	b_i (hh:mm)	ts_i (hh:mm)
0	LHU			0:00
1	HG	09:30	11:00	0:10
2	HD	09:15	11:00	0:06
3	HE	09:30	11:00	0:05
4	HF	09:15	10:30	0:10

Fig. 2 Generated route

Table 3 Van schedules

i	LH	Travel time (hh:mm)	Sevice time (hh:mm)	Time of arrival (hh:mm)	Time of leave (hh:mm)	Duration of the trip (hh:mm)
0	LHU	00:00		00:00	09:03	00:00
4	HF	00:12	00:10	09:15	09:25	00:22
3	HE	00:17	00:05	09:30	09:35	00:32
1	HG	00:22	00:10	09:40	09:50	00:47
2	HD	00:24	00:06	09:52	09:58	00:55
5	LHU	00:29		**10:03**		01:01

Table 4 New boundaries for time windows

i	health unit	a_i (hh:mm)	b_i (hh:mm)	ts_i (hh:mm)
0	LHU			0:00
1	HG	09:15	11:00	0:10
2	HD	09:00	11:00	0:06
3	HE	09:15	11:00	0:05
4	HF	09:00	10:30	0:10

With the definition of the Table 2, the following results were obtained. Figure 2 shows the generated route and Table 3 the van schedules.

The 1st solution obtained does not satisfy time limit requirement since the arrival time of samples exceeds 10:00 am. Thus a decrease in the lower limits of time windows by 15 min was decided, as presented in Table 4.

With the changed time windows, the route itself did not change. However, the total duration of the trip lessened (total of 1 h) and arrival time at HPH fitted the delivery constraints, Table 5.

Taking into account the opening hours of the LHC named HD—7:30 am—the difference in time between the first blood harvest and the delivery of samples to the

Table 5 New van schedules

i	LH	Travel time (hh:mm)	Sevice time (hh:mm)	Time of arrival (hh:mm)	Time of leave	Duration of the trip (hh:mm)
0	LHU	00:00		00:00	08:48	00:00
4	HF	00:12	00:10	09:00	09:10	00:22
3	HE	00:17	00:05	09:15	09:25	00:32
1	HG	00:22	00:10	09:30	09:40	00:47
2	HD	00:24	00:06	09:42	09:48	00:55
5	LHU	00:29		**9:53**		01:00

Table 6 Boundaries for time windows for second loop

i	Health unit	a_i (hh:mm)	b_i (hh:mm)	ts_i (hh:mm)
0	LHU			0:00
1	HG	11:00	12:00	0:10
2	HD	11:00	12:00	0:06
3	HE	11:00	12:00	0:05
4	HF	10:30	11:30	0:10

Table 7 New van schedules with waiting time

i	LH	Travel time (hh:mm)	Sevice time (hh:mm)	Time of arrival (hh:mm)	Time of leave (hh:mm)	Duration of the trip (hh:mm)	Waiting time
0	LHU	00:00		00:00	10:27	00:00	00:00
4	HF	00:12	00:10	10:39	10:49	00:22	00:00
3	HE	00:17	00:05	11:00	11:05	**00:38**	**00:06**
1	HG	00:22	00:10	11:10	11:20	00:53	00:00
2	HD	00:24	00:06	11:22	11:28	01:01	00:00
5	LHU	00:29		**11:33**		01:06	

laboratory is 2 h 23 min—higher than the 2 h (+30 min of tolerance) time window. For the second loop, in Table 6 the closing time for biological products harvesting of the health units is considered as the lower limit of the time window a_i and one hour later as the upper limit b_i.

In this case, waiting time occurs at HE since the van arrives at the point 6 min before the harvest period is over, which adds extra time to the total duration of the trip even though arrival at LHU satisfies the defined time frames (Table 7). It is possible to avoid it by increasing the lower limit of the time window for HF to 10:45 am, Table 8.

Table 8 Boundaries for time windows by increasing the lower limit of time window for HF

i	LH	Travel time (hh:mm)	Sevice time (hh:mm)	Time of arrival (hh:mm)	Time of leave (hh:mm)	Duration of the trip (hh:mm)	Waiting time
0	LHU	00:00		00:00	10:33	00:00	00:00
4	HF	00:12	00:10	**10:45**	10:55	00:22	00:00
3	HE	00:17	00:05	11:00	11:05	00:32	**00:06**
1	HG	00:22	00:10	11:10	11:20	00:53	00:00
2	HD	00:24	00:06	11:22	11:28	01:01	00:00
5	LHU	00:29		**11:33**		01:06	

The model implemented, with the time windows changed, shows that all the requirements are met and was able to provide a feasible solution taking into account the constraints of biological products handling and transportation standards.

6 Conclusion

The model was based on the existing operational requirements, which at the present require a single vehicle allocated to this biological products harvest which can only be ensured if in two loops. The solution generated by our model presents a time saving of one hour and 10 min for the 1st loop and 25 min for the second one, i.e., a total of 95 min saved for other transportation needs, which can be used for transport other products. The distribution process used by LHU was difficult to define because it was not standardized, both routes used to make the same type of deliveries and collections, regardless of the type of materials or circuit. By segregating the collection of biological products from other types of deliveries/collections, the team's awareness of the importance of the biological products arriving at the laboratories on time was raised, an issue the Service did not perceive. With regard to the indicators, the times involved in the whole route were considered, which allowed for characterization the current situation and obtaining solutions for the problem. In formulating the problem, the difficulty arose in defining the time windows, as well as the relationship between the 1st and the 2nd loop, and for this reason the problem was divided into two parts. In the first, the time windows for the 2nd loop were defined and, the gaps obtained allowed for the definition of time windows for the 1st loop. If there is to be a further collection point, the model can be adapted to ensure a suitable reduction in the route and service times. In the present case, a 95-min reduction was obtained, in a first phase, achieving savings of €2,222.64 per year. It represents a solution without any further investment or any further resources (one driver and one vehicle were kept).

As future work the problem can be improved by considering the installation of a centrifuge at a defined LHC. Considering the need to comply to all legal requirements regarding biological materials, an analysis of the financial impact of installing a new centrifuge and having an optimized 1 loop route should be done, instead of keeping the current 2 loops solution. This problem is a Pickup and Delivery problem. Although the vehicle prioritizes the collection of blood and other organic samples, as per the norm that limits their time without centrifugation, it can also pick up and deliver other programmed and urgently needed materials as it goes to the healthcare unit. The model can be improved to include parallel pickup and delivery of other items, scheduled to be collected from and delivered to other previously identified healthcare units or the hospital, as there is a capacity to do so. Despite our model having been developed to address blood collection especially, it can be improved to address other constraints in the pickup and delivery of other materials and thus address a larger scope VRP.

Acknowledgements This work is funded by the European Regional Development Fund through the Regional Operational Programme North 2020, within the scope of Project GreenHealth-Digital strategies in biological assets to improve well-being and promote green health, Norte-01-0145-FEDER-000042. The authors are grateful to the FCT Portugal for financial support by national funds FCT/MCTES to UNIAG, under Project no. UIDB/04752/2020.

References

1. Toth, P., Vigo D. (eds.): Vehicle Routing: Problems, Methods, and Applications, Second Edition No. 18 in MOS-SIAM Series on Optimization. SIAM (2014)
2. Marinakis, Y., Migdalas, A.: Annotated bibliography in vehicle routing. Oper. Res. **7**(1), 27–46 (2007)
3. Kumar, S., Panneerselvam, R.: A survey on the vehicle routing problem and its variants. Intell. Inf. Manag. **4**(3), 66–74 (2012)
4. Oliveira, J. A., Ferreira, J., Figueiredo, M., Dias, L., Pereira, G.: Sistema de Apoio á Decisão para o Transporte Não Urgente de Doentes em Veículo Partilhado, RISTI - Revista Ibérica de Sistemas e Tecnologias de Informação. 17–33 (2014)
5. Parragh, S.N., Doerner, K.F., Hartl, R.F.: A survey on pickup and delivery problems J. fur Betriebswirtschaft (2008)
6. Cordeau, J-F., Gendreau M., Laporte G., Potvin J-Y., Semet F.: A guide to vehicle routing Heuristics. J. Oper. Res. Soc. 512–522 (2002)
7. Gutierrez, A., Rocha, L.: VRP variants applicable to collecting donations and similar problems: a taxonomic review. Comput. Ind. Eng. **164** (2021)
8. Belfiore, P.P., Yoshizaki, H.T.Y.: Scatter search for heterogeneous fleet vehicle routing problems with time windows and split deliveries. Production **16**(3), 455–469 (2006)
9. Dantzig, G.B., Ramser, J.H.: the truck dispatching problem. Manag. Sci. **6**(1), 80–91 (1959)
10. Labadie, N., Christian, P., Caroline, P.: Metaheuristics Generating a Sequence of Solutions: Labadie/Metaheuristics for Vehicle Routing Problems (2016)
11. Özener, O., Ekici, A.: Managing platelet supply through improved routing of blood collection vehicles. Comput. Oper. Res. **98**, 113–126 (2018)
12. Faizal, U., Jayachitra, R., Vijayakumar, P., Rajasekar, M.: Optimization of inbound vehicle routes in the collection of bio-medical wastes, Materials Today: Proceedings, vol. 45, Part 2, pp. 692–699 (2021)

13. Cissé, M., Yalçındağ, S., Kergosien, Y., Şahin, E., Lenté, C., Matta, A.: OR problems related to Home Health Care: a review of relevant routing and scheduling problems. Oper. Res. Health Care **13–14**, 1–22 (2017)
14. Kara, I., Derya, T.: Formulations for minimizing tour duration of the traveling salesman problem with time windows. Procedia Econ. Financ. **26**, 1026–1034 (2015)

The Role of Communication on the Spread of Dengue: An Optimal Control Simulation

Artur M. C. Brito da Cruz and Helena Sofia Rodrigues

Abstract Dengue disease is a well-known disease, especially in tropical and subtropical areas. However, the communication by Health and Governmental authorities, related to personal protection and peaks of the outbreak not always is the most effective. With reliable and on-time information, people and medical staff could prepare the best response to the arrival of a new outbreak. Through a compartmental model related to vector-borne disease, with differential equations and control functional, simulations were carried out, using distinct levels of communication by authorities. The results showed that an efficient channel of communication could save money to the Health System and could considerably decrease the number of infected individuals.

Keywords Dengue · Personal protection measure · Optimal control · Skin repellent · Bed net · Insecticide-treated clothes; Communication authorities

1 Introduction

Dengue is a vector-borne disease transmitted between human hosts and usually the *Aedes aegypti* mosquito. The incidence rate of dengue fever has increased sharply worldwide, and according to Brady et al. [1] an estimated 300 million infections can occur each year. Disease symptoms range from a mild febrile illness to more severe

A. M. C. Brito da Cruz
Escola Superior de Tecnologia de Setúbal, Instituto Politécnico de Setúbal, Setúbal, Portugal
e-mail: artur.cruz@estsetubal.ips.pt

H. S. Rodrigues
Escola Superior de Ciências Empresariais, Instituto Politécnico de Viana do Castelo, Viana do Castelo, Portugal

A. M. C. Brito da Cruz · H. S. Rodrigues (✉)
Departamento de Matemática, CIDMA - Centro de Investigação e Desenvolvimento em Matemática e Aplicações, Universidade de Aveiro, Aveiro, Portugal
e-mail: sofiarodrigues@esce.ipvc.pt

J. P. Almeida et al. (eds.), *Operational Research*, Springer Proceedings in Mathematics & Statistics 411,
https://doi.org/10.1007/978-3-031-20788-4_12

symptoms such as hemorrhagic fever or even death. At this moment, patients are being treated by managing the symptoms instead of carrying out a specific treatment for the disease [9, 17, 21].

The best way to prevent infection is to avoid being bitten by mosquitoes. This way, both government and individuals could take action to diminish the mosquito population or its bites [22, 24]. From a global strategy, authorities could promote entomological surveillance to identify changes in the geographical distribution of the mosquito; in addiction, could implement vector control programs to keep vector populations at levels below a certain threshold rather than to eliminate them [15]. From an individual perspective, there are a set of personal protective measures (PPM) to reduce/eliminate mosquito bites, such as insecticide repellent, treated clothes, or treated bed nets. These measures not only help to avoid mosquito bites but also help to reduce the mosquito population by denying the blood meal essential for the nourishment of the mosquito eggs [2, 3, 11].

The information or awareness created by public health policy-makers could play a pivotal role in the decrease of a disease such as dengue. The Covid-19 pandemic shown that the implementation of any strategy should have public adherence. To raise awareness of the control measure against the mosquito, authorities should provide accurate information and disseminate it through media platforms and social networks, in order to reach distinct audience and to deflect misinformation [13].

In a global network, public health effective communication could impact the incidence of the disease, leading to new habits for the population, and at the same time, to improve the awareness of health centers' staff to prepare it logistically for an outbreak [23]. The novelty of this paper is to simulate the impact of the efficiency of policy-makers' communication on the disease development.

This way, a compartmental mathematical model was performed on Sect. 2, representing the dengue transmission. The state and control variables, as well as the parameters associated, are described. A functional used to minimize the cost of personal protective measures is added in Sect. 3. Besides, the Pontryagin maximum principle is explored to prove the existence of the optimal control solution. Section 4 is devoted to the numerical simulations, and the paper ends with the conclusion section (Sect. 5), where it summarizes the main results and carries out some future perspectives.

2 Mathematical Model

In 2012, happened the first and, at this moment, the only dengue outbreak in Madeira Island [19]. Based on the epidemiological model designed by Rodrigues et al. [19], it is proposed a new one adding a novel compartment for humans: the persons who choose to use PPM.

This model considers four state variables related to the human population and two state variables related to mosquitoes, namely:

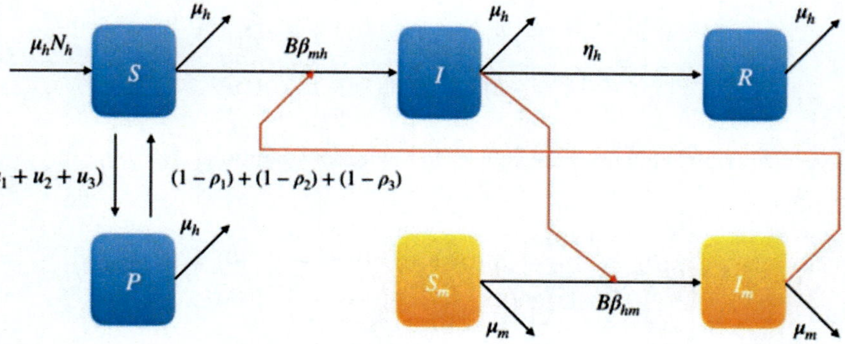

Fig. 1 Dengue epidemiological model diagram

- s – *susceptible*, individuals who can contract the disease,
- p – *protected*, individuals that use correctly protective measures,
- i – *infected*, individuals who can transmit the disease,
- r – *recovered*, individuals who have been infected and have recovered,
- s_m – *susceptible*, mosquitoes that can contract the disease,
- i_m – *infected*, mosquitoes that can transmit the disease.

In order to make a trade-off between simplicity and reality of the epidemiological model, it was considered that both humans and mosquitoes born susceptible, and there is homogeneity between host and vector populations, as well as individuals from each compartment.

Additionally, three control variables, u_1, u_2 and u_3 were added, related to the personal protection measures: skin repellent, treated bed net and insecticide-treated clothes, respectively.

An epidemiological diagram of the model is presented in Fig. 1.

The diagram depicts the mathematical model defined by a system of ordinary differential equations where it is considered that all variables are normalized:

$$
\begin{cases}
\dfrac{ds(t)}{dt} = \mu_h - (6B\beta_{mh}i_m(t) + \zeta\,(u_1(t)+u_2(t)+u_3(t)) + \mu_h)\,s(t) \\
\qquad +((1-\rho_1)+(1-\rho_2)+(1-\rho_3))p(t) \\
\dfrac{dp(t)}{dt} = \zeta\,(u_1(t)+u_2(t)+u_3(t))\,s(t) \\
\qquad -((1-\rho_1)+(1-\rho_2)+((1-\rho_3)+\mu_h)\,p(t) \\
\dfrac{di(t)}{dt} = 6B\beta_{mh}i_m(t)s(t) - (\eta_h+\mu_h)\,i(t) \\
\dfrac{dr(t)}{dt} = \eta_h i(t) - \mu_h r(t)
\end{cases}
\tag{1}
$$

and

$$\begin{cases} \dfrac{ds_m(t)}{dt} = \mu_m - (B\beta_{hm} i(t) + \mu_m) s_m(t) \\ \dfrac{di_m(t)}{dt} = B\beta_{hm} i(t) s_m(t) - \mu_m i_m(t) \end{cases} \tag{2}$$

These differential equations are subject to the initial conditions ([19]):

$$\begin{cases} s(0) = \dfrac{111991}{N_h}, p(0) = 0, i(0) = \dfrac{9}{N_h}, r(0) = 0 \\ s_m(0) = \dfrac{671000}{N_m}, i_m(0) = \dfrac{1000}{N_m}. \end{cases} \tag{3}$$

Due to its importance in the paper, the ζ parameter deserves to be highlighted. The use of controls is dependent on population knowledge of the emergence of the outbreak and the existence of protective measures to prevent/reduce the disease. Therefore, the parameter ζ describes how effectively the Health Authorities inform and persuade the population to use PPM.

The other parameters of the model, as well as their description, can be found in Table 1. Note that we have considered only the population of Funchal since the first outbreak was mainly circumscribed to this area. Some parameters were estimated or found out in other research literature, namely the ones related to mosquitoes.

In the next section is presented the functional that will be used for an optimal control approach, to simultaneously, reduce the cost associated to PPM and the number of recovered persons.

3 Optimal Control Problem

The main goal of this study is to analyze the dynamics of human state variables when using PPM, depending on the efficiency of the advertising campaigns from the Health Authorities. Moreover, those dynamics are studied when the population uses those protections accordingly with optimal control solutions. For that purpose, it is considered the control functions u_1, u_2 and u_3 that minimize the following functional

$$J(u(\cdot)) = \int_0^T \left(\gamma_1 u_1^2(t) + \gamma_2 u_2^2(t) + \gamma_3 u_3^2(t)\right) dt + R(T). \tag{4}$$

The parameters γ_1, γ_2, and γ_3 are the costs of taking personal prevention measures per day and person and the controls u_1, u_2 and u_3 represent the effort of the population of using skin repellent, bed net and insecticide-treated clothes, respectively . At the same time, the term $R(T)$, linked to the number of humans recovered by disease at the final time, $T = 365$ days, is also minimized. Note that this value represents the cumulative number of persons infected during the year. The idea of this functional

Table 1 Parameters of the epidemiological model

Parameter	Description	Range	Used values	Source
N_h	Human population		112000	[18, 19]
$\frac{1}{\mu_h}$	Average lifespan of humans (in days)		79×365	[10]
B	Average number of bites on an unprotected person (per day)		$\frac{1}{3}$	
β_{mh}	Transmission probability from I_m (per bite)	[0.25, 0.33]	0.25	[6]
$\frac{1}{\eta_h}$	Average infection period on humans (per day)	[4, 15]		[5]
$\frac{1}{\mu_m}$	Average lifespan of adult mosquitoes (in days)	[8, 45]	15	[7, 8, 14]
N_m	Mosquito population		$6 \times N_h$	[20]
β_{hm}	Transmission probability from I_h (per bite)	[0.25, 0.33]	0.25	[6]
ρ_1	Insect repellent protection (per day)		$\frac{1}{6}$	[3]
γ_1	Insect repellent cost (per person and day)		$\frac{10 \times 12}{365 \times 112000}$	[3]
ρ_2	Bed net protection (per day)		$\frac{1}{3}$	[3]
γ_2	Bed net cost (per person and day)		$\frac{20}{365 \times 112000}$	[3]
ρ_3	Insecticide-treated clothes protection (per day)		$\frac{1}{2}$	[3]
γ_3	Insecticide-treated clothes cost (per person and day)		$\frac{30 \times 6}{365 \times 112000}$	[3]

is twofold: reduce the individual costs to prevent infection, and minimize the total number of recovered persons at the final time.

Rigorously, the aim is to find the optimal control values u_1^*, u_2^* and u_3^* that minimizes the objective functional and such that the state trajectories $s^*, p^*, i^*, r^*, s_m^*, i_m^*$ are solutions of the Eqs. (1) and (2) with the following initial conditions:

$$s(0) \geqslant 0, \quad p(0) \geqslant 0, \quad i(0) \geqslant 0, \quad r(0) \geqslant 0, \quad s_m(0) \geqslant 0, \quad i_m(0) \geqslant 0 \tag{5}$$

and within the set of admissible controls

$$\Omega = \{u_i(\cdot) \in L^{\infty}[0, T] : 0 \leqslant u_i(\cdot) < 1, \forall t \in [0, 365], i = 1, 2, 3\}.$$

The cost function J is L^2 since the integrand function is convex with respect to the controls u_1, u_2 and u_3. Furthermore, systems (1) and (2) are Lipschitz with respect to the state variables and, therefore, exists an optimal control [4].

The Hamiltonian function is defined by

$$\begin{aligned}
H = H(s(t), p(t), i(t), r(t), s_m(t), i_m(t), \Lambda, u(t)) &= \gamma_1 u_1^2(t) + \gamma_2 u_2^2(t) + \gamma_3 u_3^2(t) \\
&+ \lambda_1 (\mu_h - (6B\beta_{mh} i_m(t) + \zeta (u_1(t) + u_2(t) + u_3(t)) + \mu_h) s(t)) \\
&+ \lambda_1 ((1 - \rho_1) + (1 - \rho_2) + (1 - \rho_3)) p(t) \\
&+ \lambda_2 (\zeta (u_1(t) + u_2(t) + u_3(t))) s(t) - ((1 - \rho_1) + (1 - \rho_2) \\
&+ (1 - \rho_3) + \mu_h p(t)) \\
&+ \lambda_3 (6B\beta_{mh} i_m(t) s(t) - (\eta_h + \mu_h) i(t)) \\
&+ \lambda_4 (\eta_h i(t) - \mu_h r(t)) \\
&+ \lambda_5 (\mu_m - (B\beta_{hm} i(t) + \mu_m) s_m(t)) \\
&+ \lambda_6 (B\beta_{hm} i(t) s_m(t) - \mu_m i_m(t))
\end{aligned}$$

Pontryagin's Maximum Principle [16] states that there exists a nontrivial absolutely continuous mapping, the adjoint vector:

$$\Lambda : [0, 365] \to \mathbb{R}^6, \ \Lambda(t) = (\lambda_1(t), \lambda_2(t), \lambda_3(t), \lambda_4(t), \lambda_5(t), \lambda_6(t))$$

such that

$$s' = \frac{\partial H}{\partial \lambda_1}, \quad p' = \frac{\partial H}{\partial \lambda_2}, \quad i' = \frac{\partial H}{\partial \lambda_3}, \quad r' = \frac{\partial H}{\partial \lambda_4}, \quad s_m' = \frac{\partial H}{\partial \lambda_5} \text{ and } i_m' = \frac{\partial H}{\partial \lambda_6},$$

where the optimality condition

$$\begin{aligned}
&H\left(s^*(t), p^*(t), i^*(t), r^*(t), s_m^*(t), i_m^*(t), \Lambda(t), u^*(t)\right) = \\
&= \min_{0 \leqslant u_1, u_2, u_3 \leqslant 1} H\left(s^*(t), p^*(t), i^*(t), r^*(t), s_m^*(t), i_m^*(t), \Lambda(t), u(t)\right)
\end{aligned}$$

and the transversality conditions

$$\lambda_i(365) = 0, \quad i = 1, 2, 3, 5, 6 \quad and \quad \lambda_4(365) = 1 \tag{6}$$

hold almost everywhere in [0, 365].

Applying Pontryagin's maximum principle, we obtain the following result.

Theorem 1 *The optimal control problem with fixed final time $T = 365$ defined by the Eqs. (1)–(4) has a unique solution $(s^*(t), p^*(t), i^*(t), r^*(t), s_m^*(t), i_m^*(t))$ with the adjoint function satisfying*

$$\begin{cases} \lambda_1'(t) = & \lambda_1\left(6B\beta_{mh}i_m^*(t) + \zeta\left(u_1^*(t) + u_2^*(t) + u_3^*(t)\right) + \mu_h\right) - \\ & -\lambda_2\zeta\left(u_1^*(t) + u_2^*(t) + u_3^*(t)\right) - \lambda_3 6B\beta_{mh}i_m^*(t) \\ \lambda_2'(t) = & -\lambda_1\left((1-\rho_1) + (1-\rho_2) + (1-\rho_3)\right) + \lambda_2((1-\rho_1) + (1-\rho_2) \\ & +(1-\rho_3)) \\ \lambda_3'(t) = & \lambda_3\left(\eta_h + \mu_h\right) - \lambda_4\eta_h + \left(\lambda_5 - \lambda_6\right)B\beta_{hm}s_m^*(t) \\ \lambda_4'(t) = & \lambda_4\mu_h \\ \lambda_5'(t) = & \lambda_5\left(B\beta_{hm}i^*(t) + \mu_m\right) - \lambda_6 B\beta_{hm}i^*(t) \\ \lambda_6'(t) = & \left(\lambda_1 - \lambda_3\right)6B\beta_{mh}s^*(t) + \lambda_6\mu_m \end{cases} \tag{7}$$

and with the optimal controls $u_1^*(\cdots)$, $u_2^*(\cdots)$ and $u_3^*(\cdots)$ on $[0, T]$ given by

$$\begin{cases} u_1^*(t) = \max\left\{0, \min\left\{\dfrac{\zeta\left(\lambda_1(t) - \lambda_2(t)\right)s^*(t)}{2\gamma_1}, 1\right\}\right\} \\ u_2^*(t) = \max\left\{0, \min\left\{\dfrac{\zeta\left(\lambda_1(t) - \lambda_2(t)\right)s^*(t)}{2\gamma_2}, 1\right\}\right\} \\ u_3^*(t) = \max\left\{0, \min\left\{\dfrac{\zeta\left(\lambda_1(t) - \lambda_2(t)\right)s^*(t)}{2\gamma_3}, 1\right\}\right\} \end{cases}.$$

4 Numerical Results

The results were obtained by implementing the problem on MATLAB version R2017b and it was used a forward-backward fourth-order Runge-Kutta method with a variable time step for efficient computation (see [12] for more details).

Four different scenarios were designed to simulate the efficiency of the communication of the Health Authorities. Parameter ζ measures that efficiency and the values 25%, 50%, 75%, and 100% are considered and reflect the amount of population informed and, at the same time, strictly follows the prescribed individual protective measures. To make a fair comparison, a naive scenario where the population is not informed and does not take any kind of protection is also reproduced.

The different types of protection used (insect repellent, bed nets, and insecticide-treated clothing) have a different time of protection, due to their characteristics. For example, while each application of insect repellent lasts 4 hours, the bed net is considered to protect 8 hours a day. Table 1 displays the price of each protection/control (parameter γ) and also the corresponding time of application/duration (parameter ρ). Furthermore, for distinct ζ, various simulations were carried out: no control, optimal control using only one control per time, and optimal control approach using the three controls simultaneously.

4.1 25% of the Population Willing to Take Protective Measures

Informing only a quarter of the population that an outbreak is occurring can make a big difference in the dynamics of human state variables related to the disease (see Fig. 2). In this scenario, the predictions showed a reduction of 15%, when using insect repellent, up to 73%, when using all controls combined, of the total number of infected persons.

When a single control is used, regardless of which one, the results obtained are similar and not particularly effective. However, the use of all controls combined

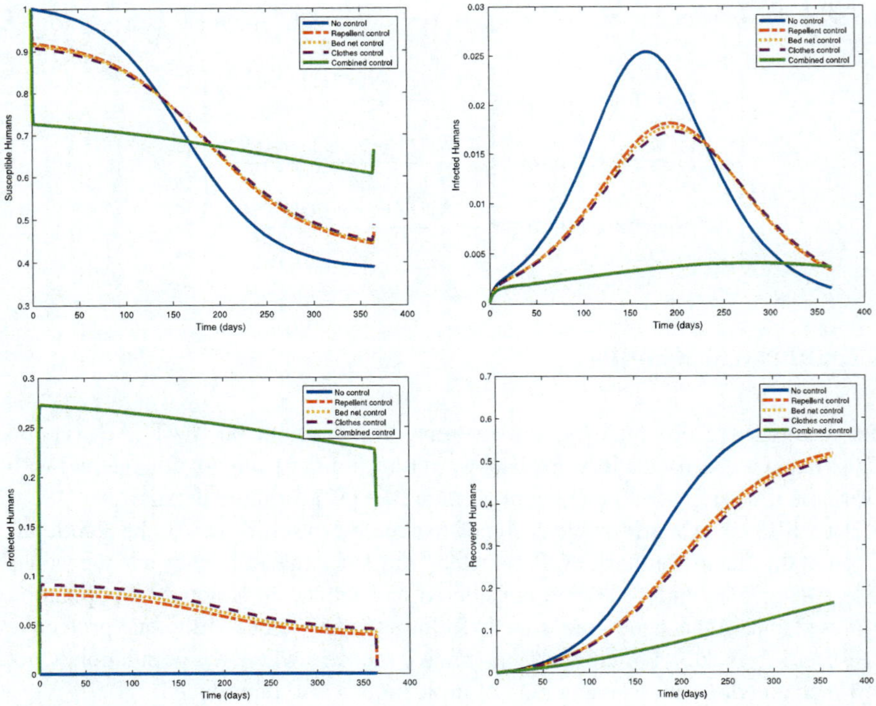

Fig. 2 25% population informed—Human state variables

Table 2 25% population informed—Summary of the simulations

Strategy	Peak of infected persons	Peak's day	$R(T)$	Money spent
No control	2843	168	68393	0
Only insect repellent control	2024	192	57818	119.7
Only bed net control	1979	195	57087	20
Only insecticide-treated clothes	1927	194	56249	179.5
Combined control	447	263	18321	319.4

drops the total number of infected people to less than 16% of the population. Table 2 lists the maximum number of infected people in one day, and on what day that happened (peak day). Also on the table, we have the number of recovered persons at the final time, and the cost, on average, that each protected individual will spend for buying personal equipment during the whole year. The amount of money that each protected person spend per year is given by

$$\int_0^T \gamma_1 u_1(t) + \gamma_2 u_2(t) + \gamma_3 u_3(t) dt.$$

With an effective communication reaching out 25% of the population, and using a single control, the human state dynamics do not significantly differentiate from the no control scenario. However, the combined control approach not only have a significant impact on the number of infected persons, but also the peak's day of infection is much later than the other cases.

4.2 *50% of the Population Willing to Take Protective Measures*

Half the population taking protective measures insures that almost three quarters of them will not be infected when is used a single control and more than 97% will be protected when in use of the three controls. In Fig. 3 it can be seen that the infected persons' curve is being flatten compared to the 25% curve. This means that the peak day of infection is happening sooner, but with fewer infected persons, and the end of the epidemic is happening later proved by simulations done over a large period of time.

Naturally, on Table 3 and due to the fact that more people is using PPM, the peak of infected persons is lower in each situation that is used protection. Although the

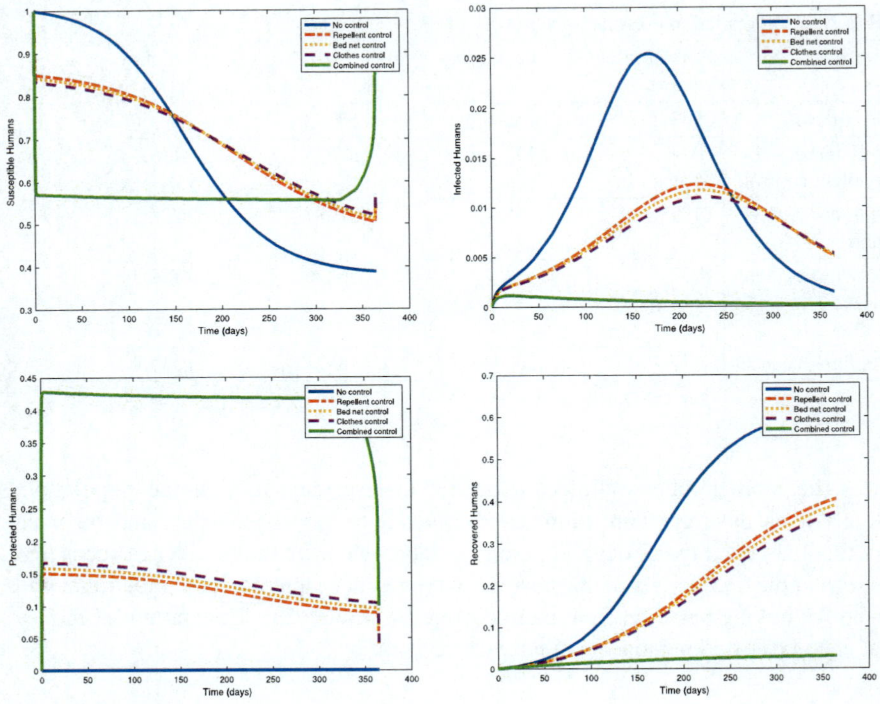

Fig. 3 50% population informed—Human state variables

Table 3 50% population informed—Summary of the simulations

Strategy	Peak of infected persons	Peak's day	$R(T)$	Money spent
No control	2843	168	68393	0
Only Insect repellent control	1375	219	45235	119.9
Only Bed net control	1305	222	43563	20
Only insecticide-treated clothes	1229	229	41660	179.8
Combined control	132	16	3285	310.4

outbreak is not finished after a year, this reveals that the control measures take effect in the fight against the disease.

As expected, all the combined controls cost decreases due to the fact that the outbreak is less severe.

4.3 75% of the Population Willing to Take Protective Measures

With most of the population taking protective measures, it would be expected to have a small number of infected persons. While this is true when all the controls are used, when is used a single control the results are marginally better in comparison with the 50% of people informed and have the willingness to protect themself (see Fig. 4).

Surely, this is a better solution because the total number infected people drops (slightly) but also because flattens the infected curve helping health organizations to better manage the crisis.

The information observed in the graphics meets the expectation of the results in Table 4.

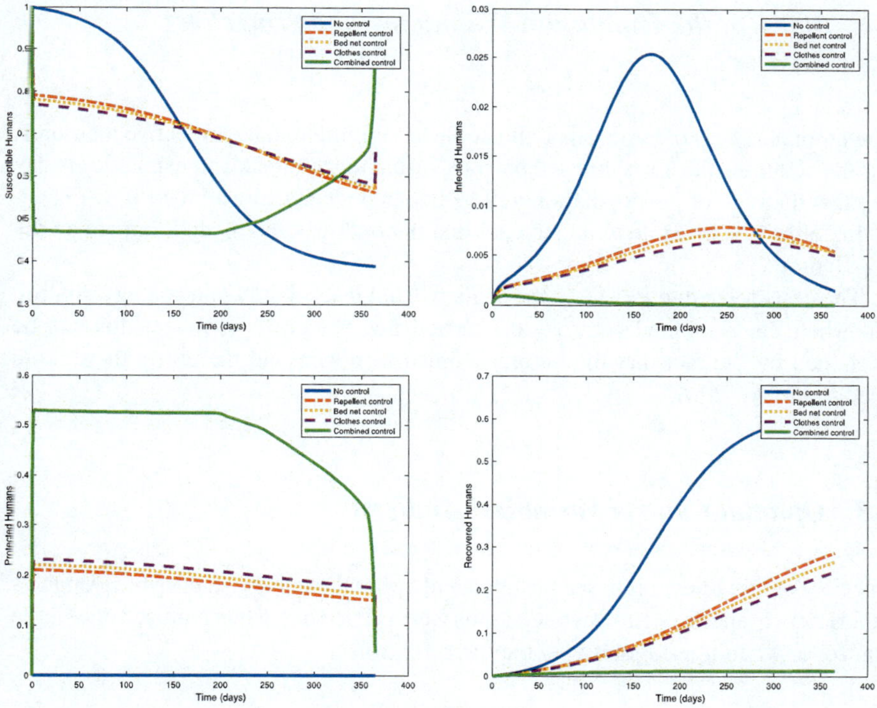

Fig. 4 75% population informed—Human state variables

Table 4 75% population informed—Summary of the simulations

Strategy	Peak of infected persons	Peak's day	$R(T)$	Money spent
No control	2843	168	68393	0
Only Insect repellent control	878	248	31936	119.9
Only Bed net control	800	251	29584	20
Only insecticide-treated clothes	718	258	27030	179.8
Optimal control	103	12	1388	261.1

4.4 100% of the Population Willing to Take Protective Measures

The utopian scenario because not all people are willing to take protective measures, has the better results. Less infected persons during the outbreak and in one single day and less than 1% of the population will be infected when using all the PPM (Fig. 5).

In Table 5, is possible to understand that the outbreak has a small impact on the population.

The costs related to PPM, are the same when it is used only one strategy/control, but when the combined controls are carried out, the cost decreases; this can be explained by the fact that the outbreak finishes sooner, and therefore, there is no need to use the PPM.

4.5 Optimal Control Variables' Analysis

The core of this study it is to see the impact of optimal control on a possible Dengue's outbreak in Madeira Island. Two scenarios were carried out, when it is used one single control or when it used jointly all the three controls.

4.5.1 Single Control

In the previous tables, it was shown that if the communication of the outbreak reaches more people, costs associated with using only one PPM keep getting higher. Figure 6 shows that the use of a single control is not effective, reason why each control has the needs to be used all the year, even with different levels of the population being minded to use protection. This could leave to a fatigue situation from using protective measures, similar to what has been seen with Covid-19.

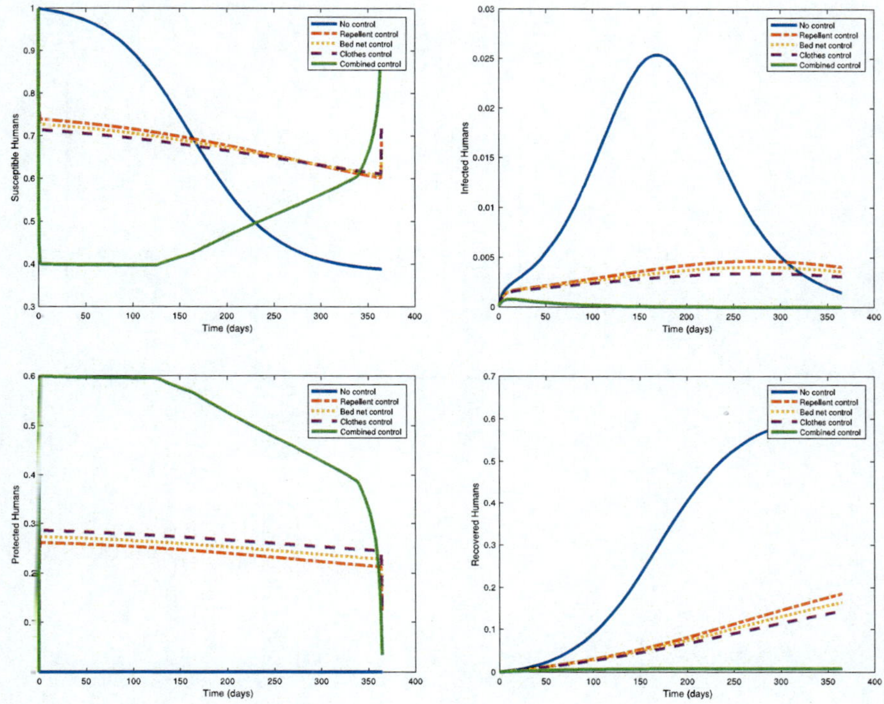

Fig. 5 100% population informed—Human state variables

Table 5 100% population informed—Summary of the simulations

Strategy	Peak of infected persons	Peak's day	$R(T)$	Money spent
No control	2843	168	68393	0
Only insect repellent control	191	42	20590	119.9
Only bed net control	184	38	18300	20
Only insecticide-treated clothes	177	34	15986	179.8
Optimal control	86	12	865	213.9

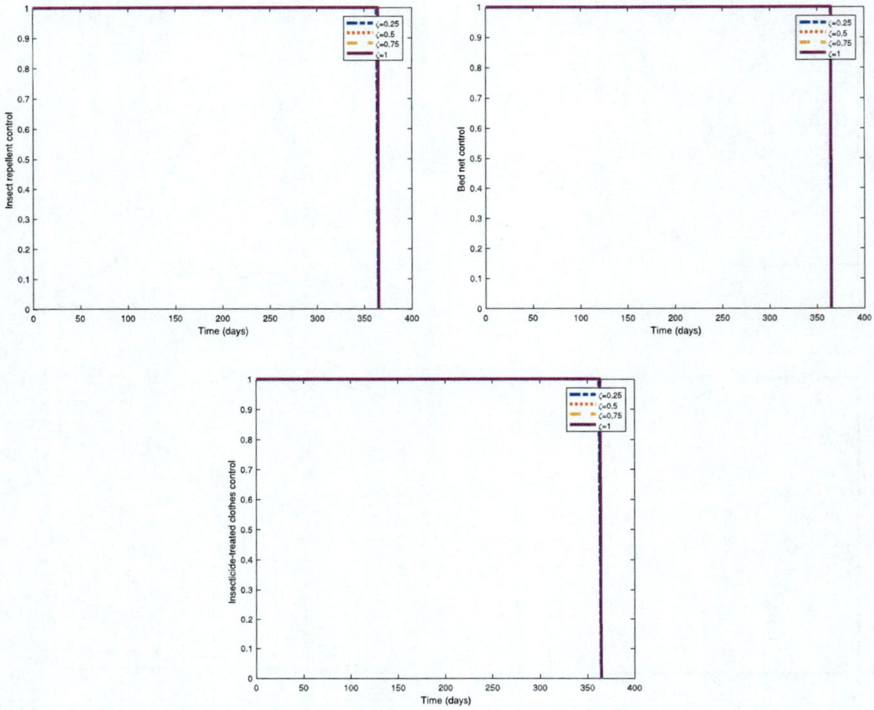

Fig. 6 Optimal control using only one protection

4.5.2 Combined Control

Using three controls at the same time gives better protection to each person throughout the day. Moreover, if more people use them, then fewer people get infected and the outbreak ends sooner. Again, while bed net should be used most of the year, other controls drop from maximum control value much sooner due to its costs. Figure 7 shows that when more people take immediate protective measures, due to efficient communication from Health Authorities, the sooner those restrictions can be lightened and, therefore, costs associated decrease.

5 Conclusions

This research has analyzed the importance of effective communication by health and government authorities during an outbreak of dengue. Accurate information related to protective measures and the level of incidence could contribute to the flattening of the curve disease. In a situation of no control, meaning that no one takes any PPM, the total of infected individuals reaches $R(T) = 68393$. As the effectiveness

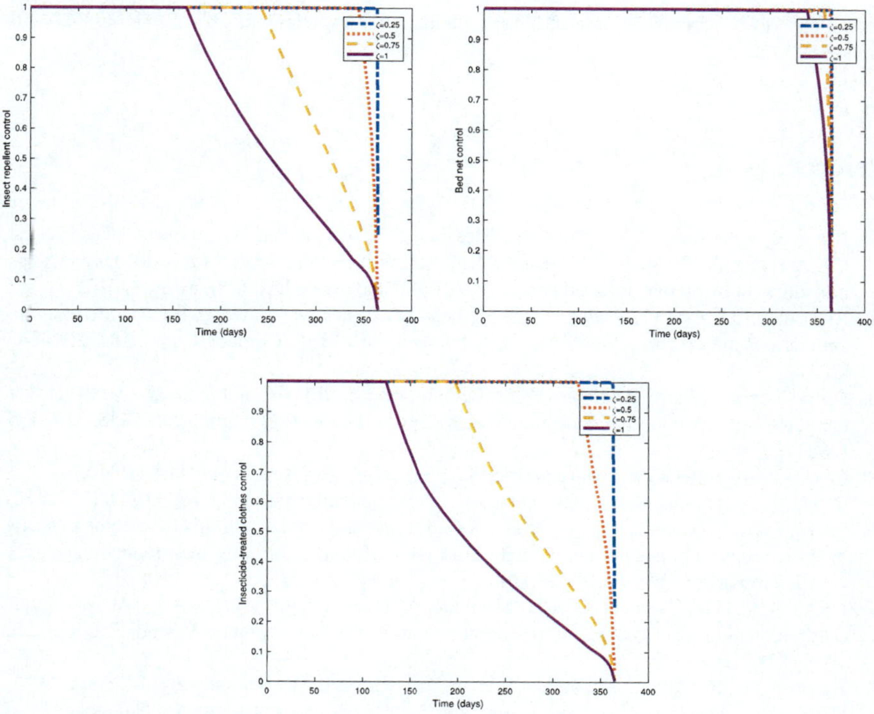

Fig. 7 Optimal control using combined protections

of communication increases (ζ), the curve of infected individuals decreases sharply. In addition, analyzing the recovered curve in the combined control of PPM, it passes from 18321 individuals recovered when $\zeta = 0.25$, to 865 persons affected by the disease when $\zeta = 1$.

Regardless the effectiveness of the communication, the behavior of the state curves (susceptibles, protected, infected, and recovered) using controls separately are similar (Sects. 4.1 to 4.4). The notorious change happens when the combined controls are adopted, leading to the rise of people that decide to protect themselves and, consequently, decline dramatically the number of infected persons.

Analyzing each control separately (Sect. 4.5), it is observed the need to use all time the control because is not efficient to fight the disease. However, when the same control is combined with others, the time of its application reduces, since the outbreak tends to end earlier.

In future work, the cost of the official communication also should be taken in consideration in the optimal control problem, to understand the burden of disease.

Acknowledgements This work is supported by The Center for Research and Development in Mathematics and Applications (CIDMA) through the Portuguese Foundation for Science and

Technology (FCT—Fundação para a Ciência e a Tecnologia), references UIDB/04106/2020 and UIDP/04106/2020.

References

1. Brady, O.J., Gething, P.W., Bhatt, S., Messina, J.P., Brownstein, J.S., Hoen, A.G., Moyes, C.L., Farlow, A.W., Scott, T.W., Hay, S.I.: Refining the global spatial limits of dengue virus transmission by evidence-based consensus. PLoS Negl. Trop. Dis. **6**(8), e1760 (2012)
2. Brito da Cruz, A.M.C., Rodrigues, H.S.: Personal protective strategies for dengue disease: simulations in two coexisting virus serotypes scenarios. Math. Comput. Simul. **188**, 254–267 (2021)
3. Brito da Cruz, A.M.C., Rodrigues, H.S.: Economic burden of personal protective strategies for dengue disease: an optimal control approach. Optim. Learn. Alg. Appl. CCIS **1488**, 319–335 (2021)
4. Cesari, L.: Optimization-Theory and Applications. Springer, New York, USA (1983)
5. Chan, M., Johansson, M.A.: The incubation periods of dengue viruses. PLoS One **7**(11) (2012)
6. Focks, D.A., Brenner, R.J., Hayes, J., Daniels, E.: Transmission thresholds for dengue in terms of Aedes aegypti pupae per person with discussion of their utility in source reduction efforts. Am. J. Trop. Med. Hyg. **62**, 11–18 (2000)
7. Focks, D.A., Haile, D.G., Daniels, E., Mount, G.A.: Dynamic life table model for Aedes aegypti (Diptera: Culicidae): analysis of the literature and model development. J. Med. Entomol. **30**, 1003–1017 (1993)
8. Harrington, L.C., Buonaccorsi, J.P., Edman, J.D., Costero, A., Kittayapong, P., Clark, G.G., Scott, T.W.: Analysis of survival of young and old Aedes aegypti (Diptera: Culicidae) from Puerto Rico and Thailand. J. Med. Entomol. **38**, 537–547 (2001)
9. Hung, T.M., Clapham, H.E., Bettis, A.A., Cuong, H.Q., Thwaites, G.E., Wills, B.A., Boni, M.F., Turner, H.C.: The estimates of the health and economic burden of dengue in Vietnam. Trends Parasitol. **34**(10), 904–918 (2018)
10. INE, Statistics Portugal. http://censos.ine.pt. Accessed 5 Apr. 2021
11. Kroeger, A., Ordonez-Gonzalez, J., Behrend, M., Alvarez, G.: Bednet impregnation for Chagas disease control: a newperspective. Trop. Med. Int. Health **4**, 194–198 (1999)
12. Lenhart, C.J., Workman, J.T.: Optimal Control Applied to Biological Models. Chapman & Hall/CRC, Boca Raton, FL, USA (2017)
13. Liu, N., Chen, Z., Bao, G.: Role of media coverage in mitigating COVID-19 transmission: Evidence from China. Technol. Forecast. Soc. Change **163**, No. 120435 (2021)
14. Maciel-de-Freitas, R., Marques, W.A., Peres, R.C., Cunha, S.P., Lourenço-de-Oliveira, R.: Variation in Aedes aegypti (Diptera: Culicidae) container productivity in a slum and a suburban district of Rio de Janeiro during dry and wet seasons. Mem. Inst. Oswaldo Cruz **102**, 489–496 (2007)
15. Pan American Health Organization: Technical Document for the Implementation of Interventions Based on Generic Operational Scenarios for Aedes Aegypti Control. PAHO, Washington, D.C. (2019)
16. Pontryagin, L.S., Boltyanskii, V.G., Gamkrelidze, R.V., Mishechenko, E.F.: The Mathematical Theory of Optimal Processes VIII + 360. Wiley, New York/London (1962)
17. Pulkki-Brännström, A.M., Wolff, C., Brännström, N., Skordis-Worrall, J.(2012). Cost and cost effectiveness of long-lasting insecticide-treated bed nets—a model-based analysis. Cost Effect. Res. Alloc. **10**(5), 1–13
18. Rocha, F.P., Rodrigues, H.S., Monteiro, M.T.T., Torres, D.F.M.: Coexistence of two dengue virus serotypes and forecasting for Madeira Island. Oper. Res. Health Care **7**, 122–131 (2015)
19. Rodrigues, H.S., Monteiro, M.T., Torres, D.F.M., Silva, A.C., Sousa, C.: Dengue in Madeira Island. In: Bourguignon, J.P., Jeltsch, R., Pinto, A., Viana, M. (eds.) International Conference

and Advanced School Planet Earth, DGS II, CIM Series in Mathematical Sciences, vol. 1, pp. 593–605. Springer, Cham (2015)
20. Rodrigues, H.S., Monteiro, M.T.T., Torres, D.F.M., Zinober, A.: Dengue disease, basic reproduction number and control. Int. J. Comput. Math. **89**(3), 334–346 (2012)
21. Shepard, D., Undurraga, E.A., Halasa, Y., Stanaway, J.D.: The global economic burden of dengue: a systematic analysis. Lancet—Infect. Dis. **16**(8), 935–941 (2016)
22. World Health Organization: Managing Regional Public Goods for Health: Community-Based Dengue Vector Control. Asian Development Bank and World Health Organization, Philippines (2013)
23. World Health Organization: Keeping the vector out: housing improvements for vector control and sustainable development. Geneva: World Health Organization; 2017. Licence: CC BY-NC-SA 3.0 IGO (2017)
24. World Health Organization: Global vector control response 2017–2030. Geneva: World Health Organization; 2017. Licence: CC BY-NC-SA 3.0 IGO (2017)

Towards an Optimized and Socio-Economic Blood Supply Chain Network

Ana Torrado and Ana Paula Barbosa-Póvoa

Abstract In this work, a two-stage allocation model involving healthcare facilities with blood services is developed to design the Blood Supply Chain (BSC), where economic and social aspects were considered. In the first stage, the design of a BSC network is considered to support blood supply and demand, and the geographical distribution for donors/patients according to the location of the healthcare facilities. Based on the first stage results, the product flow among blood centers (BC) and hospitals, as well as the minimization of costs are studied in the second stage. Economic aspects were considered through cost minimization while the social aspect was explored by allocating donors/patients to the closest facilities. Exploratory experiments are conducted using Portuguese National Health Services data to test the model's applicability. From this, it was concluded that there is a need for additional blood services for the collection phase, and a large number of healthcare facilities with non-licensed blood services should be licensed in the considered SC network. Regarding donors, the allocation costs represent 90% of the total costs, meaning that more types of collection facilities are needed in the context of our study. For patients, adding healthcare facilities with licensed blood services represents the higher costs (78%). Concerning the product flow optimization, the production costs correspond to 82%. Additionally, the model allows the improvement of the distribution of the hospitals according to the existing BCs at reduced costs.

Keywords Blood supply chain · Socio-Economic · MILP · Optimization

A. Torrado (✉) · A. P. Barbosa-Póvoa
Center for Management Studies, Instituto Superior Técnico, University of Lisbon, Lisbon, Portugal
e-mail: ana.torrado@tecnico.ulisboa.pt

A. P. Barbosa-Póvoa
e-mail: apovoa@tecnico.ulisboa.pt

J. P. Almeida et al. (eds.), *Operational Research*, Springer Proceedings in Mathematics & Statistics 411,
https://doi.org/10.1007/978-3-031-20788-4_13

1 Introduction

Blood is scarce and has a prominent role in human life. As the availability of blood products through healthcare services is often a matter of life or death to the patient, blood shortages or outdated blood products should not occur when managing the Blood Supply Chain (BSC). As a result, managing the Supply Chain (SC) decisions to deal with minimum unmet demand levels is crucial [1]. The BSC network plays an important role in the healthcare systems all over the world [2], presenting different activities and services, such as blood collection, testing, processing, inventory, distribution, and transfusion in hospitals [3, 4]. As a service-oriented SC, BSC includes critical features—perishability, the freshness of transfused blood, multiplicity and respective lifetimes of blood products, fluctuations in blood supply and demand, and blood compatibility among donors and patients—involving several cooperating stakeholders [5]. These features increase the complexity of such systems and consequently, it is important to design and plan accordingly to them [6, 7]. During the last years, Blood Supply Chain Network Design (BSCND) and planning assumed a significant role in promoting the increased performance of these SCs while capturing the researcher's attention [6, 8, 9]. BSCND focuses on designing the SC's network concerning the system's entities (e.g., number of collection facilities, storage facilities, transportation vehicles) and the respective improvement of the processes involved. Additionally, reaching an efficient BSCND is a strategy to follow, that should contribute to refining the SC performance through the coordination of the different processes to guarantee the optimal match between blood supply and demand [10, 11]. Any improvement in the management of these SCs will positively impact the supply of blood as a life-saving product [2]. The location-allocation decisions involved (e.g., related to the number of infrastructures and respective allocation of donors to the collection infrastructures) are one example of possible improvement which can contribute to the achievement of an optimal BSCND. An optimal BSCND solution must guarantee collection strategies, production factors, social and geographical considerations [6, 12] that support high quality at the lowest cost and in the shortest time. This is challenging and implies accurate management and a sustainable mindset (while considering economic, environmental, and social dimensions) [13]. First, the economic dimension involves the trade-off of minimizing total costs and maximizing social benefits [14]. Second, the concept of going green corresponds to a necessity nowadays [15]. Third, concerning the social dimension, according to the World Health Organization (WHO), improving the health of populations, responsiveness, and fairness/equity are defined as three main purposes for health systems [7]. Based on the combination of BSCND and sustainability, this work aims to contribute to improvements in the BSC. However, this is a first step of our research that explores the social and economic dimensions of sustainability and aims to easily support the decision-makers when designing and planning the BSC. Thus, this work does not consider perishability and uncertainty, important aspects of the problem in study, but considers assumptions related to the minimum demand and supply that should be taken into account allowing a buffer to deal with uncertainty in these critical variables. It takes into consideration: (1) the estimated percentage of the population that will provide or receive blood, (2) the lowest distances that contribute to better geographical accessibility and equity, and (3) the respective related transportation and production costs, as well as (4) the analysis of the product flow optimization that contributes to understanding the best distribution of hospitals about the available blood centers (BC). To reach that aim, a simple two-stage network model for the design and plan of the BSC is developed and proposed. It considers the strategic design of the BSC, simultaneously with the tactical planning. This model incorporates the collection, production in the BCs, and transfusion in the hospitals. It defines (i) the design of a BSC network that regulates the fluctuations in supplies and demands while considering the geographical distribution for donors/patients, and (ii) the optimal product flow among the BCs and hospitals. The developed work is validated in an exploratory case study to show the respective applicability.

The remainder of this work is structured as follows. Section 2 presents an overview of the existing literature related to this work. In Sect. 3, a comprehensive description of the problem is given, as well as, in Sect. 4, the characterization of the model formulation is explained. Section 5 presents some input data used in the case study. Section 6 demonstrates the applicability of the model. Finally, conclusions and future research directions are drawn in Sect. 7.

2 Literature Review

Generally, the BSC location-allocation decisions follow a strategic/tactical planning nature, involving not only the SC entities but also the donors that donate blood at the collection infrastructures and the patients that will receive that donated blood at the demand nodes. In this section, the location-allocation studies related to the BSC, and connected with social and economic sustainability dimensions, are analyzed.

2.1 *Social Dimension*

Concerning the social pillar, Ramezanian and Behboodi [8] concentrated on the increase of blood donors' utility to reduce shortages, but also on strategies to motivate donors to donate blood (such as distance of blood donors from blood facilities, the experience of donors, and advertising budget). Focusing on the SC strategies, in particular on the minimization of the distances, Karadağ [5] analyses the re-organization of a BSC to generate improvements, presenting a novel multi-objective mixed-integer location-allocation model for a BSC design problem, consisting of mobile and permanent units. There are other aspects to take into consideration that influence the performance of the BSC, such as uncertainty of demand, geographical accessibility, prioritizing patients based on urgency levels, and hierarchical structure of networks. Zarrinpoor [16] proposed a novel reliable hierarchical location-allocation model, addressing a real-world health service network design problem, which considers the previous key issues. Additionally, also Gilani Larimi and Yaghoubi [17] reported main issues that affect the quantity of blood for donation, in particular different types of donors (first-time, regular), the number of booked/non-booked donors within and after working hours, the submission of social announcements, and the allocation of blood extraction technologies to the demand nodes. However, the technology allocation in hospitals is expensive, but technologies can support the hospitals' efficiency.

2.2 Economic Dimension

Concentrating on the economic pillar, Nagurney and Masoumi [18] developed a network design model for a BSC to determine the optimal network capacities under demand uncertainty of perishable products, considering costs (discarding, shortages/surpluses at the demand points), and quantifying the supply-side risk associated with procurement. Nagurney [19] developed a generalized network optimization model for regional blood banking systems handling demand uncertainty, including collection sites, testing, processing facilities, storage facilities, distribution centers, and demand points. Zahiri [2] presented a mixed-integer linear programming (MILP) model to make strategic-tactical decisions in a blood collection system over a multi-period planning horizon. Zahiri and Pishvaee [20] addressed the design of the BSC network, adding the blood group compatibility. A bi-objective mathematical programming model is developed to minimize costs and unsatisfied demand. Attari and Jami [21] concentrated in expand a regionalized BC system with just one blood product, aiming to minimize the total costs (of establishing and relocating facilities, operational and delivering blood) and to minimize the average delivery time among facilities. Osorio [12, 22, 29] presented different models which support the decision-makers to redesign the SC network, by considering the blood products at different levels of centralization intending the determination of the ideal collection and production strategies to minimize the total costs. Hamdan and Diabat [23] presented a two-stage stochastic programming problem, considering eight distinct blood types, to achieve the optimal BSC network able to minimize the outdated blood units applying First In First Out (FIFO) network costs, and blood delivery time. They determine the number of mobile blood collection facilities, as well as inventory and production decisions. Samani [24] proposed a multilateral perspective for BSC network design, using a novel multi-objective mathematical model by incorporating both quantitative and qualitative factors—aiming to minimize the loss of product freshness and total cost of the SC. Reza [25] proposed a bi-objective model for an integrated blood SC network design to minimize the total network cost and maximize the quality factor. On the other hand, Bruno [26] formulated the facility location problem and aimed, at reorganizing regional blood management systems, to reduce total management costs without compromising the self-sufficiency goal. Arani [9] studied a BSC network design consisting of donors, blood collection facilities, BCs, and hospitals while considering the ABO-Rh factors and shelf lives of blood products. An integrated inventory system for sharing hospitals' inventory levels (lateral resupply) was used to investigate cross-matching and outdated units—to satisfy the demand by the other hospitals' inventories in the absence of the required product at the BC and the excess in any hospital. There are also studies joining location-allocation and inventory problems [27, 28]. Hsieh [27] studied a two-echelon SC in which each regional BC sends blood to different BCs and then delivers it to different allocated hospital blood banks. First, the model is proposed to obtain the location-allocation decisions, by determining how many BCs should be in an area, where BCs should be located, and which services should be assigned to Community Blood Centers (CBCs). Second, a model is implemented to acquire the inventory control decisions to achieve optimal blood replenishment quantity for each CBC. Both models aimed to minimize the total SC costs and the maximization of the responsiveness level. Hosseini-Motlagh [28] developed a bi-objective two-stage stochastic programming model to determine the optimum location-allocation and inventory management decisions; aiming to minimize the total cost of the SC—fixed, operating, inventory holding costs, wastage, and transportation costs—along with minimizing the substitution levels to provide a safe transfusion.

From the above analysis it can be said that several works have been published considering strategic decisions, however, few have a social vision—while considering the demand and the supply of blood at the same time in a multi-product decision approach. Additionally, it is important to notice that the environmental aspects are relevant to completing the sustainability approach in BSC [30]. In this work, the environmental dimension will be indirectly explored, by reducing the traveled distances and consequently the associated emissions. The main aim of this study is to propose a two-stage socio-economic BSC model that provides a strategic-tactical plan for designing a sustainable BSC, addressing the problems related to the mismatch between supply and demand of blood concerning an expected population of donors and patients in a location, but also the respective minimization of costs for donors and patients. SC's costs, such as production and transportation costs to measure the SC's performance in terms of economic goals are considered. Regarding the social component, the distribution of the population among the different hospitals and BCs in Portugal is included to pursue a more equitable BSC (in terms of geographical equity and geographical access measured by population density and traveled distance).

The next sections are dedicated to the problem definition and mathematical formulation, taking into account the previous goals in mind.

3 Problem Definition

The location-allocation of blood facilities affects the utility for donors and patients. The location of the BSC facilities usually results in weak geographical accessibility. Improving the accessibility of demand zones while considering the equality of demand satisfaction is vital. Blood donations occur via main or temporary facilities. Focusing on fixed facilities, establishing them is more costly. However, main fixed facilities have higher levels of capacity and technology allowing to decrease possible risks to the BSC. The received whole blood is sent to BCs to be separated into the blood components (platelets; red blood cells, RBCs; and plasma). In this study, the BSC network corresponds to a single-product (whole blood) network before the BCs; and after the BC, the network transforms into a multi-product BSC. All the perishable blood components are sent from BCs to the demand zones. The allocation problem is divided into two stages and solved sequentially (observe Fig. 1). In the first stage, the usage of the existent SC superstructure (healthcare facilities) to deal with the reception and delivery of blood is considered and the necessary capacity optimized,with respect to supply (donors) and demand (patients). The goal is to find how existent licensed facilities should be used and which non-licensed facilities should be licensed to accommodate donors and patients from different districts (locations). The donors provide the whole blood in the healthcare facilities and the patients can receive different blood products. In the second stage, the healthcare facilities (for patients) obtained from the previous stage are considered and the optimal flow/ allocation of the RBCs (A+/−, B+/−, AB+/−, O+/−), among hospitals and BCs, is performed while minimizing production and transportation costs. Concerning this

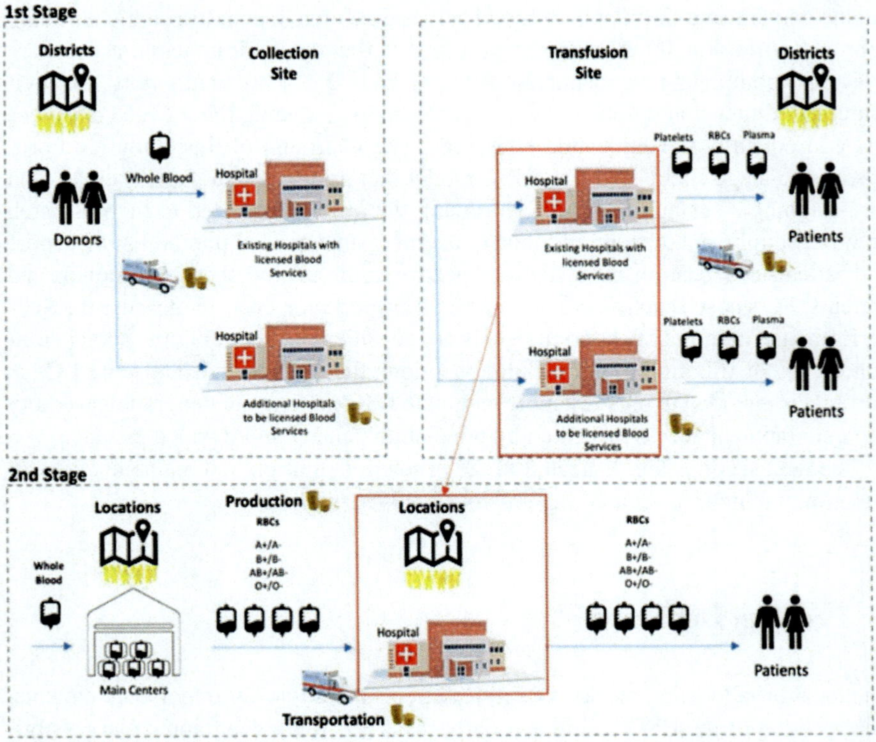

Fig. 1 BSC network and respective relationship between the two stages

stage, the transformation of whole blood in RBCs or transshipment among hospitals or BCs is not considered.

In this problem the following assumptions are considered:

(1) a single product (whole blood) is shipped through the network before processing (into multiple products);
(2) the multiple products are transported from the BCs to the Hospitals, after the processing phase;
(3) at each facility of the network (BCs and hospitals) the existing and the potential capacity is limited;
(4) the distance between the entities of the SC is considered as a direct distance;
(5) each district has licensed facilities, but also has the option of adding licensing to the existing facilities to increase the overall capacity to handle the blood need/offer.

As stated, the problem above is solved through a two-stage approach involving in each stage the solution of a MILP model. The first stage model is a simple facility-location model that informs decisions on (1) how to best use the existent capacity, with respect to donors and patients; (2) whether to add new licenses to non-licensed facilities and with what capacity; and (3) how patients/donors from a district should be allocated to the various facilities to ensure that the facilities can provide treatment for the patients and to maximize the demand/supply coverage. No transformation processes within the facilities are considered. Concerning the second stage, the distribution network defined in stage 1 is explored aiming to find the optimal flow of blood products given the transportation and production policies.

4 Mathematical Formulation

The two-stage model is described below. The sets allow the definition of the network structure; and each level of the SC is defined by an entity (hospitals working as collection sites, BCs, hospitals operating as transfusion sites).

Indices and sets

d, index of the different districts, $d \in D$;
h, index of the existing healthcare facilities with licensed blood services, $h \in H$;
a, index of the additional healthcare facilities (without existent licensed blood services) to be added in the network, $a \in A$;
f, index of the total (existing and additional) healthcare facilities, $f \in F = HUA$;
p, index of the blood products, $p \in P$
bc, index of the locations of blood centers, $bc \in BC$
bt, index of the blood type products, $bt \in BT$

Parameters

First stage:
$dem_{d,p}$, Expected population in district d that will need blood product p;
$sup_{d,p}$, Expected population in district d who will provide whole blood p;
$cap_{f,p}$, Population served by a healthcare facility f that need blood product p;
$travDist_{d,f}$, Traveled distance between district d and healthcare facility f;
$travCost_{d,f}$, Cost according to traveled distance between district d and healthcare facility f;
$adCost_a$, Cost of adding licensed blood services a;
$pltyCap$, Penalty of adding extra capacity

Second stage:
$qnt_{bc,f,bt}$, Quantity (per units) of type of blood product bt produced and transported from a blood center bc to a hospital f in a regular (weekly planned) time;
$transpCost_{bt}$, Transportation cost per unit of a specific bt;
$travDist_{bc,f}$, Traveled distance between blood center bc and hospital f;
$dem_{f,bt}$, Demand at hospital h of type blood product bt units;
$capProd_{bc,bt}$, Capacity of the blood center bc to produce blood product bt;
$prodCost_{bt}$, Production cost for a specific type of blood product bt when considering a regular (weekly planned) time;
$reghCap_{bc,bt}$, Regular hour capacity at blood center bc per blood type bt;
$perProdDistr_{bt,bc}$, Percentage of production and distribution of type of blood product bt placed in blood center bc

Decision Variables

First stage:
$adF_{a,p}$, equal to 1 if we license facility a (with non-licensed blood services) to provide/receive a product p; 0, otherwise;
$adCap_{a,p}$,Extra capacity of new licensed facility a to provide product p;
$pop_{d,f,p}$, Population from district d served by a facility f and by product p;

Second stage:
$prodProd_{bc,bt}$, equal to 1 if a blood center bc produces that type of blood product type bt, else, 0

Two objective functions are presented, the first objective function for stage 1 and the second objective function for stage 2. The first objective function (Eq. 1) corresponds to the minimization of costs for donors and patients. The traveled distances and geographical distribution as social costs are considered in the first stage. The first term corresponds to the traveling cost from the different districts to the healthcare facilities; the second term corresponds to the costs of licensing existing healthcare facilities for providing blood services; and the third term corresponds to the cost of adding extra capacity to the new licensed facilities.

$$MinCosts = \sum_{d \in D} \sum_{f \in F} pop_{d,f,p}.travDist_{d,f}.travCost_{d,f} + \sum_{a \in A} adF_{a,p}.adCost_a + pltyCap.\sum_{a \in A} adCap_{a,p} \tag{1}$$

The second objective function (Eq. 2) aims to find the product flow from the BCs to the hospitals. The first term provides the production cost and the second term gives the transportation cost.

$$MinCosts(Prod. + Transp.) = \sum_{bc \in BC} \sum_{f \in F} \sum_{bt \in BT} qnt_{bc,f,bt}.(prodCost_{bt} + (transpCost_{bt}.travDist_{bc,f})).prodProd_{bc,bt} \tag{2}$$

In both stages, the geographical equity is included through the variation in geographical utilization and geographical access measured by density and traveled distance in terms of the population. The equity of access corresponds to the minimization of the total travel time, and consequently leads to less environmental impacts related to emissions, by ensuring that individuals that provide blood or require blood transfusion receive the blood needed at the closest available service; as well as, by guaranteeing that the access to facilities is being provided to as many individuals as possible. The geographical equity minimizes the level of unmet need for the geographical area(s) with the highest level of unmet need. In this way, it will allow the maximum allocation of blood in the geographical area(s) with the lowest blood provision.

The constraints of this exploratory study involved in the first stage model are dedicated to the relation among the expected population of a district that will need blood concerning the respective estimated demand (1% of the resident population of a district) (3); the expected population of a district that will provide blood according to the respective estimated supply (2% of the resident population of a district)—that should be higher than the demanded blood (4). Additionally, the authors added as constraints the capacity of a blood facility that should satisfy the expected population of a district (5); and that the expected population of a district should be satisfied, if needed with additional blood facilities or extra capacity (6).

$$\sum_{f \in F} pop_{d,f,p} = dem_{d,p}, \forall d \in D, p \in P\ , \tag{3}$$

$$dem_{d,p} \leq sup_{d,p}, \forall d \in D, p \in P\ , \tag{4}$$

$$\sum_{d \in D} pop_{d,h,p} \leq cap_{h,p}, \forall h \in H, p \in P\ , \tag{5}$$

$$\sum_{d \in D} pop_{d,a,p} \leq cap_{a,p}.adF_{a,p} + adCap_{a,p}, \forall a \in A,\ p \in P\ , \tag{6}$$

Concerning the second-stage model, this has as input the healthcare facilities defined in stage 1 (f). New constraints are added corresponding to the demand satisfaction at the hospital level (7); the production capacity and product constraint (8); production rate constraint (9) and non-negativity constraints (10).

$$\sum_{f \in F} qnt_{bc,f,bt} = dem_{f,bt}, \forall bc, bt \tag{7}$$

$$\sum_{f \in F} qnt_{bc,f,bt} \leq prodCost_{bc}.capProd_{bc,bt}, \forall bc, bt \tag{8}$$

$$\sum_{f \in F} qnt_{bc,f,bt} \leq reghCap_{bc,bt}.perProdDistr_{bc,bt}.prodProd_{bc,bt}, \forall bc, bt \tag{9}$$

$$\begin{aligned} &dem_{d,p}, sup_{d,p}, cap_{f,p}, travDist_{d,f}, travCost_{d,f}, adCost_a, qnt_{bc,f,bt}, \\ &\quad transpCost_{bt}, travDist_{bc,f}, dem_{f,bt}, capProd_{bc,bt}, prodCost_{bt}, \\ &\qquad reghCap_{bc,bt}, perProdDistr_{bt,bc} \geq 0 \end{aligned} \tag{10}$$

5 Case Study

A Portuguese based BSC is studied (Fig. 2).

Data related to the existing licensed blood services and possible blood services (that can be added in the network), for donors and patients, was considered (NHS of Portugal, "Imunohemoterapia", from 2017). Figure 2 presents the representation of the locations of healthcare facilities for patients that receive blood, donors that provide blood and the location of each BC. According to the previous formulation, the healthcare facilities (f) include hospitals with existing and licensed blood services (h), and existing facilities (a) that may be licensed with blood services. Regarding the donors, 19 existing healthcare facilities (with licensed blood services) were considered and 4 healthcare facilities with non-licensed blood services were proposed; and for the patients, 24 existing healthcare facilities (with licensed blood services) were added in the model and 19 healthcare facilities with non-licensed blood services were proposed.

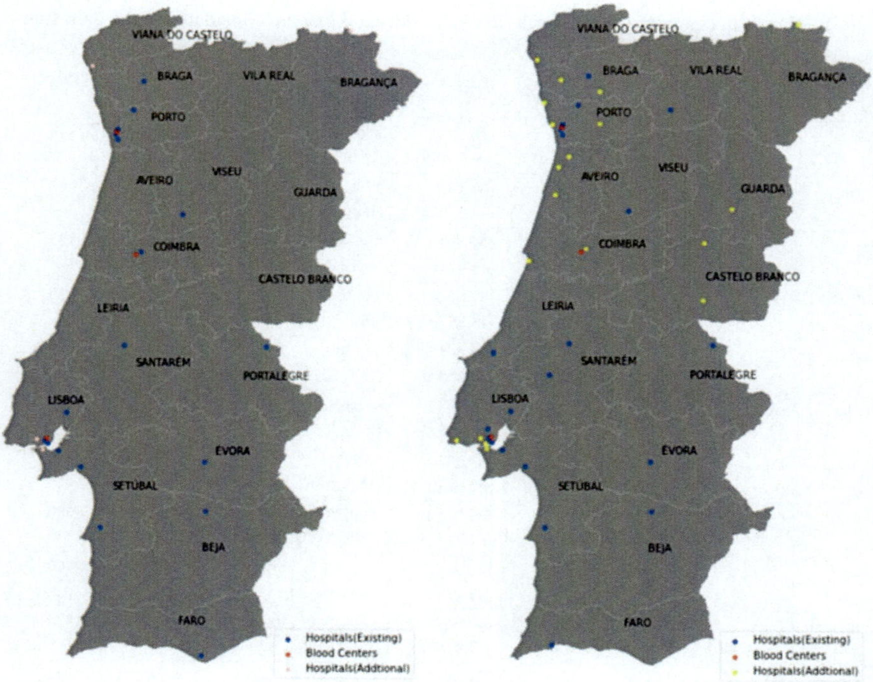

Fig. 2 Healthcare Facilities (including hospitals with existing and licensed blood services, possible non-licensed blood services to be (additionally) licensed in the network and BCs: for donors provide blood (left); for patients receive blood (right)

Table 1 contains the characteristics of each district in Portugal, namely region, the area of each district, coordinates, resident population, population density, as well as, the assumptions for demand (considering 1% of the resident population) and supply (2% of the resident population).

6 Computational Experiments

The above two-stage procedure was modeled in Python and solved through Gurobi on a 2.3 GHz Intel Core i9 and 16 GB of RAM computer. It is important to note that the results below correspond to an exploratory study as the model developed is simple. So these results can only be viewed as a first attempt towards the solution of a more complex BSC. The results were achieved by first solving the first stage and the obtained results were then used as input to solve the second stage of the problem. In the results, three different aspects are analyzed: social aspects; costs; and product flow optimization.

Table 1 The main characteristics of each district in Portugal that are considered in the case study

Region	District	District ID	Area (km^2)	Coordinates	Resident population (hab.)	Population density (hab./km^2)	Estimated demand	Estimated supply
LVT	D1	Lisboa	2761	(38.7071, −9.13549)	2863433 (INE,2019)	821	28634	57268
North	D2	Porto	2395	(41.15, −8.61024)	1778146 (INE,2018)	742	17781	35562
North	D3	Braga	2706	(41.5518, −8.4229)	956185 (INE,2011)	313	9561	19123
LVT	D4	Setúbal	5064	(38.5245, −8.89307)	911794 (INE,2009)	171	9117	18235
North	D5	Aveiro	2798,54	(40.6412, −8.65362)	714200 (INE,2011)	262	7142	14284
Center	D6	Leiria	3505,78	(39.7443, −8.80725)	470895 (INE,2011)	134	4708	9417
LVT	D7	Santarém	6747	(39.2362, −8.68707)	465701 (INE,2009)	70	4657	9314
Algarve	D8	Faro	4960	(37.0154, −7.93511)	434023 (INE,2009)	87,5	4340	8680
Center	D9	Coimbra	3947	(40.2115, −8.4292)	429987 (INE,2011)	109	4299	8599
Center	D10	Viseu	5007	(40.6575, −7.91428)	391215 (INE,2011)	78,13	3912	7824
North	D11	Viana do Castelo	319,02	(41.6946, -8.83016)	88725 (INE,2011)	278,1	887	1774
North	D12	Vila Real	4328	(41.2959, −7.74635)	213775 (INE,2011)		2137	4275
Center	D13	Castelo Branco	6675	(39.8239, −7.49189)	196264 (INE,2011)	34	1962	3925
Center	D14	Guarda	5518	(40.5371, −7.26785)	168898 (INE,2011)	30,61	1688	3377
Alentejo	D15	Évora	7393	(38.571, −7.9096)	168034 (INE,2009)	23	1680	3360
Alentejo	D16	Beja	10229,05	(38.0156, −7.86523)	152758 (INE,2011)	15	1527	3055
North	D17	Bragança	6608	(41.8072, −6.75919)	136252 (INE,2011)	21	1362	2725
Alentejo	D18	Portalegre	6065	(39.2914, −7.43235)	118506 (INE,2011)	19,5	1185	2370

6.1 Social Component Analysis

The results from the first stage are presented in Figs. 3 and 4 and Tables 2 and 3. The social component analysis, associated with the geographical distribution of donors and patients among healthcare facilities, is essentially analyzed in stage 1. Figure 3 represents the number of healthcare facilities needed by the district in view of the whole blood donation. The obtained distribution related to the number of donors,

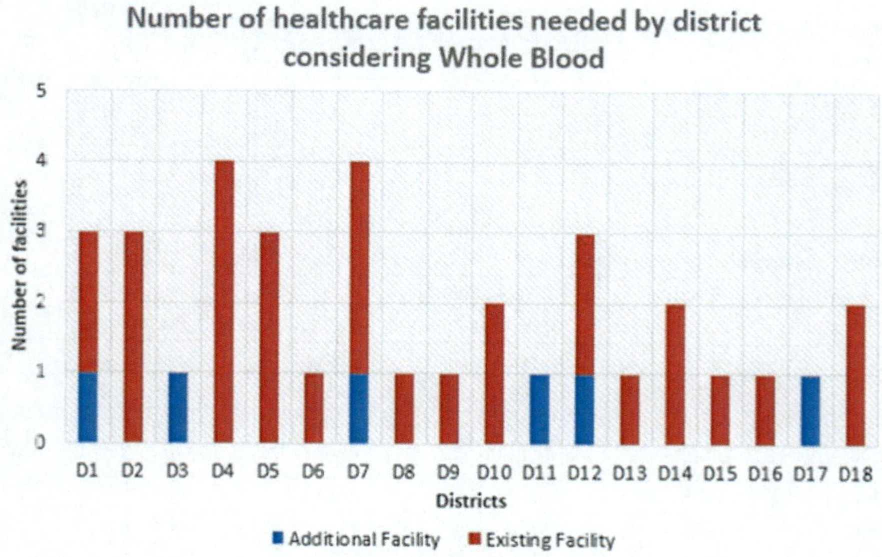

Fig. 3 Number of healthcare facilities needed by district considering the whole blood

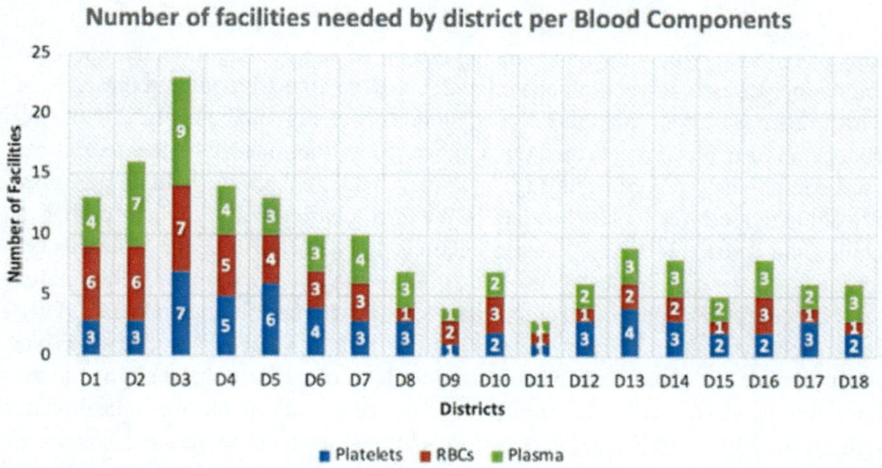

Fig. 4 Number of healthcare facilities (with licensed and non-licensed blood services) needed by district per blood product. Each facility is able to provide more than one blood product

for example for Lisbon (D1), considering healthcare facilities as unique infrastructures (type) to receive donors is not completely operational. The model identified 3 facilities that will receive 8636 (F3 with licensed blood services), 5481 (F14 with licensed blood services), and 37421 (F20 with non-licensed blood services) donors in Lisbon (Table 2). Regarding the SC network, all the healthcare facilities with non-licensed blood services are required (in the SC network) and should be licensed (in

Table 2 Distribution of donors (D1)

Region	Facility	No. Donors (for WB)
LVT	F3	8636
LVT	F14	5481
LVT	F20	37421

Table 3 Distribution of patients (D1)

Region	Facility	No. Patients		
		Platelets	RBCs	Plasma
LVT	F4	1465	7525	1356
LVT	F5	902	5300	1639
LVT	F16	N.A.	1660	N.A.
LVT	F18	762	1667	180
LVT	F28	N.A.	2320	95
LVT	F34	N.A.	3759	N.A.
LVT	F37	N.A.	N.A.	N.A.

terms of blood services) to overcome the estimated supply. Additionally, and concerning the patients, to accommodate the demand, the model identified that at least 18 healthcare facilities with non-licensed blood services should be licensed in the SC network. Figure 4 has represented the number of facilities needed by the district per blood component. According to Fig. 4, district 3 (D3, Braga) has a higher number of healthcare facilities (corresponding to the third district with a higher number of people) and district 11 (D11, Viana do Castelo) has a lower number of healthcare facilities (and people) allocated for the different blood components. Additionally, it should be noticed that one facility can provide different blood products. Focusing on Lisbon (D1) and taking into account the Table 3, this district that has the higher number of people, and presents the distribution of the products aligned to the population needs, 4 healthcare facilities are needed to provide plasma, 6 healthcare facilities for RBCs and 3 healthcare facilities for platelets—however, these facilities are also able to deliver additional blood products to patients. Regarding the optimized patients' distribution of the different blood components for Lisbon (D1): (1) for platelets, the number of patients to be transfused by year is 1465, 902, and 762 for each selected healthcare facility; (2) for plasma, the number of patients by year to be transfused is around 1356, 1639, 180, and 95 (F28 with non-licensed blood services) for each selected healthcare facility; and (3) for the RBCs, the number of patients to be transfused by year is around 7525, 5300, 1660, 1667, 2320 (F28 with non-licensed blood services) and 3759 (F34 with non-licensed blood services) for each selected healthcare facility.

6.2 Costs Component Analysis

Concerning the costs related to patients, the RBCs components are the most transfused product, as a result, RBCs will involve more costs in adding healthcare facilities with licensed blood services (78% for RBCs), as well as, in allocating the patients to the existing licensed healthcare facilities (65% for RBCs). Regarding the costs associated with the donors, allocating them (according to the population distribution) will involve 90% of the costs, while additional licensed blood services (in healthcare facilities) correspond to 10% of the costs. This result shows that there is a need for more infrastructures for collection services to reduce the allocation costs.

6.3 Product Flow Optimization

In the second stage, the focus is on the product flow that was analyzed for just one product (the RBCs), while considering the same healthcare facilities obtained from stage 1 (for patients) and adding the BCs facilities in the SC network. The following distribution of healthcare facilities and demands for each BC was found: BC1 should provide blood to 16 healthcare facilities (where demand corresponds to 51%); BC2 should be responsible for 14 facilities (where demand corresponds to 26%) and BC3 should manage blood to 8 facilities (demand of 23%) (Figs. 5 and 6). According to the results, production costs denote 82% and transportation costs correspond to 18% of the total costs.

In this exploratory study, it is assumed in the beginning more than 40 healthcare facilities (in stage 1) satisfy the demand (of patients); however, in stage 2, focusing

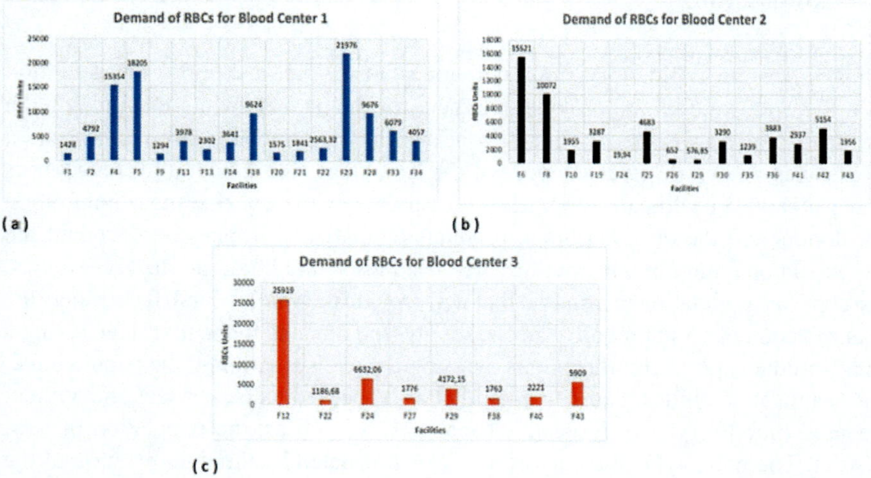

Fig. 5 Product Flow Optimization: Demand of RBCs for (a) BC 1, (b) BSC 2 and (c) BC 3

Fig. 6 Representation of healthcare facilities and demand among the existent BCs distributed in Portugal

just on the RBCs product flow optimization just 38 healthcare were needed. As stated before the model developed is just a first attempt to approach the problem. Future developments are required to represent additional complexities related to the design of the BSC.

7 Conclusions

This work is an exploratory study that appears as a first attempt to the definition of a BSC network able to provide treatment for the patients while maximizing the demand/supply coverage. A simple location-allocation model is formulated and a two stages procedure is developed. In the first stage, the model allocates donors and patients to healthcare facilities (by districts), while considering whole blood for donors and the different blood products for patients. It takes into account the expected population in a district that will need a specific blood product, as well as, the expected population in a district that will provide the whole blood. It considers the minimization of costs for both aspects, supply, and demand in the first stage (using a multi-product approach). It explores the geographic distribution of the population to the healthcare facilities (based on facilities with licensed blood services and without licensed blood services), respective capacities, as well as, the population of each district. The geographical equity of access is translated by the minimization of the total travel time by ensuring that individuals that provide blood or require blood transfusion receive the blood needed at the closest available service. In this way, it will allow the maximum allocation of blood in the geographical areas with the lowest blood provision. On the other hand, in the second stage, the model is able to optimize the blood units allocation (assuming the healthcare facilities obtained in the first stage) to the BCs based on the RBCs consumption.

Concerning sustainability, (i) BSC costs, such as production and transportation costs are considered to measure the SC's performance in terms of economic goals; and, (2) the social component is explored based on the distribution of the population among the different healthcare facilities to pursue a more equitable BSC. In this work, the authors do not just look at the donors or patients, but we aim to find the product flow from the BCs and the hospitals—in a specific case study—looking for clusters of hospitals allocated to each specific BC. By achieving the blood product flow optimization, we were able to determine how to cluster the hospitals to the BCs—based on population (distribution) needs, adding production aspects in the model, as well as, the transportation part among the BCs and hospitals.

This initial model approach can be the basis to support the decision-makers, since we join the geographical utilization and the geographical access measured by density and traveled distance in terms of the population, and at the same time connecting it with the collection, production, distribution, and transfusion stages.

In future improvement, the model developed should be integrated and enhanced by (i) adding the processing, inventory, and transshipment operations; (ii) having the contribution to the environmental dimension approach; (iii) introducing the dynamics of the problem (e.g. time), and (iv) as a result of the multiple uncertainties that affect the BSC, applying stochastic or robust optimization approaches could be used to solve the problem.

References

1. Ghorashi, S.B., Hamedi, M., Sadeghian, R.: Modeling and optimization of a reliable blood supply chain network in crisis considering blood compatibility using MOGWO. Neural Comput. Appl. **32**, 12173–12200 (2020). https://doi.org/10.1007/s00521-019-04343-1
2. Zahiri, B., Torabi, S.A., Mousazadeh, M., Mansouri, S.A.: Blood collection management: Methodology and application. Appl. Math. Model. **39**, 7680–7696 (2013). https://doi.org/10.1016/j.apm.2015.04.028
3. Beliën, J., Forcé, H.: Supply chain management of blood products: a literature review. Eur. J. Oper. Res. **217**, 1–16 (2012). https://doi.org/10.1016/j.ejor.2011.05.026
4. Osorio, A.F., Brailsford, S.C., Smith, H.K.: A structured review of quantitative models in the blood supply chain: a taxonomic framework for decision-making. Int. J. Product. Res. **53**, 7191–7212 (2015). https://doi.org/10.1080/00207543.2015.1005766
5. Karadağ, İ, Keskin, M.E., Yiğit, V.: Re-design of a blood supply chain organization with mobile units. Soft Comput. **25**, 6311–6327 (2021). https://doi.org/10.1007/s00500-021-05618-3
6. Samani, M.R.G., Hosseini-Motlagh, S.M.: An enhanced procedure for managing blood supply chain under disruptions and uncertainties. Ann. Oper. Res. **283**, 1413–1462 (2019). https://doi.org/10.1007/s10479-018-2873-4

7. Haeri, A., Hosseini-Motlagh, S.M., Ghatreh Samani, M.R., Rezaei, M.: A mixed resilient-efficient approach toward blood supply chain network design. Int. Trans. Oper. Res. **27**, 1962–2001 (2020). https://doi.org/10.1111/itor.12714
8. Ramezanian, R., Behboodi, Z.: Blood supply chain network design under uncertainties in supply and demand considering social aspects. Transp. Res. Part E: Logist. Transp. Rev. **104**, 69–82 (2017). https://doi.org/10.1016/j.tre.2017.06.004
9. Arani, M., Chan, Y., Liu, X., Momenitabar, M.: A lateral resupply blood supply chain network design under uncertainties. Appl. Math. Modell. **93**, 165–187 (2021). https://doi.org/10.1016/j.apm.2020.12.010
10. Fanoodi, B., Malmir, B., Jahantigh, F.F.: Reducing demand uncertainty in the platelet supply chain through artificial neural networks and ARIMA models. Comput. Biol. Med. **113** (2019). https://doi.org/10.1016/j.compbiomed.2019.103415
11. Hosseini-Motlagh, S.M., Samani, M.R.G., Cheraghi, S.: Robust and stable flexible blood supply chain network design under motivational initiatives. Socio-Economic Plan. Sci. **70** (2020). https://doi.org/10.1016/j.seps.2019.07.001
12. Osorio, A.F., Brailsford, S.C., Smith, H.K., Blake, J.: Designing the blood supply chain: how much, how and where? Vox Sanguinis **113**, 760–769 (2018). https://doi.org/10.1111/vox.12706
13. Moslemi, S., Pasandideh, S.H.R.: A location-allocation model for quality-based blood supply chain under IER uncertainty. RAIRO—Oper. Res. **55**, S967–S998 (2021). https://doi.org/10.1051/ro/2020035
14. Banini, J., Lim, M.K., Lim, M, Anosike, A.: LITERATURE ANALYSIS IN SUSTAINABLE SUPPLY CHAIN MANAGEMENT: LEADERSHIP STYLE AND CULTURE Integration of Lean Six Sigma and Green View project New Journal: Clean Technologies and Recycling View project (2021)
15. Jemai, J., Chung, B., Sarkar, B.: Environmental effect for a complex green supply-chain management to control waste: a sustainable approach. J. Clean. Prod. **277**, 122919 (2020)
16. Zarrinpoor, N., Fallahnezhad, M.S., Pishvaee, M.S.: The design of a reliable and robust hierarchical health service network using an accelerated Benders decomposition algorithm. Eur. J. Oper. Res. **265**, 1013–1032 (2018). https://doi.org/10.1016/j.ejor.2017.08.023
17. Gilani Larimi, N., Yaghoubi,S.: A robust mathematical model for platelet supply chain considering social announcements and blood extraction technologies. Comput. Ind. Eng. **137** (2019). https://doi.org/10.1016/j.cie.2019.106014
18. Nagurney, A., Masoumi, A.H.: Supply chain network design of a sustainable blood banking system. In: International Series in Operations Research and Management Science, , pp. 49–72. Springer, New York LLC (2012)
19. Nagurney, A., Masoumi, A.H., Yu, M.: Supply chain network operations management of a blood banking system with cost and risk minimization. Comput. Manag. Sci. **9**, 205–231 (2012). https://doi.org/10.1007/s10287-011-0133-z
20. Zahiri, B., Pishvaee, M.S.: Blood supply chain network design considering blood group compatibility under uncertainty. Int. J. Prod. Res. **55**, 2013–2033 (2017). https://doi.org/10.1080/00207543.2016.1262563
21. Attari, M.Y.N., Jami, E.N.: Robust stochastic multi-choice goal programming for blood collection and distribution problem with real application. J. Intell. Fuzzy Syst. **35**, 2015–2033 (2018). https://doi.org/10.3233/JIFS-17179
22. Osorio, A. F., Brailsford, S.C., Smith, H.K., Forero-Matiz, S.P.: Simulation-Optimization model for production planning in the blood supply chain. Health Care Manag. Sci. **20**(4), 548–64 (2017). https://doi.org/10.1111/vox.12706
23. Hamdan, B., Diabat, A.: A two-stage multi-echelon stochastic blood supply chain problem. Comput. Oper. Res. **101**, 130–143 (2019). https://doi.org/10.1016/j.cor.2018.09.001
24. Samani, M.R.G., Hosseini-Motlagh, S.M., Ghannadpour, S.F.: A multilateral perspective towards blood network design in an uncertain environment: methodology and implementation. Comput. Ind. Eng. **130**, 450–471 (2019). https://doi.org/10.1016/j.cie.2019.02.049
25. Reza, M., Samani, G., Hosseini-Motlagh, S.-M., Sheshkol, M.I., Shetab-Boushehri, S.-N.: A bi-objective integrated model for the uncertain blood network design with raising products quality (2019)

26. Bruno, G., Diglio, A., Piccolo, C., Cannavacciuolo, L.: Territorial reorganization of regional blood management systems: evidences from an Italian case study. Omega (UK) **89**, 54–70 (2019). https://doi.org/10.1016/j.omega.2018.09.006
27. Hsieh, C.L.: An evolutionary-based optimization for a multi-objective blood banking supply chain model. In: Modern Advances in Applied Intelligence. Lecture Notes in Computer Science, vol. 8481. Springer, Cham (2014). https://doi.org/10.1007/978-3-319-07455-9_53
28. Hosseini-Motlagh, S.M., Samani, M.R.G., Homaei, S.: Blood supply chain management: robust optimization, disruption risk, and blood group compatibility (a real-life case). J. Ambient Intell. Hum. Comput. **11**, 1085–1104 (2020). https://doi.org/10.1007/s12652-019-01315-0
29. Osorio, A.F., Brailsford, S.C., Smith, H.K.: Whole blood or apheresis donations? A multi-objective stochastic optimization approach. Eur. J. Oper. Res. **266**(1), Elsevier B.V., pp. 193–204 (2018). https://doi.org/10.1016/j.ejor.2017.09.005
30. Torrado, A., Barbosa-Póvoa, A.: Towards an optimized and sustainable blood supply chain network under uncertainty: a literature review. Clean. Logist. Supply Chain. **3** (2022). https://doi.org/10.1016/j.clscn.2022.100028

A DEA Approach to Evaluate the Performance of the Electric Mobility Deployment in European Countries

Clara B. Vaz and Ângela P. Ferreira

Abstract This work aims to assess the performance of European countries on the deployment of low-emission vehicles in road transportation. For this purpose, a model based on Data Envelopment Analysis (DEA) is used to calculate a composite indicator for several European countries, aggregating seven sub-indicators built from a data set for the 2019 year. Various virtual weight restrictions schemes of the sub-indicators are studied to explore the robustness of the performance results. By adopting the most robust scheme, the performance results obtained indicate that most European countries have the potential to improve their practices towards better road transport sustainability, by emulating the best practices observed in the four identified benchmarks. Thus, the inefficient countries should take measures to better support the market share of plug-in electric vehicles. In addition, the railway sector and the penetration of renewable energies should be enhanced to improve road transportation sustainability.

Keywords DEA · Road sustainability · Composite indicator

C. B. Vaz (✉) · Â. P. Ferreira
Research Centre in Digitalization and Intelligent Robotics (CeDRI), Instituto Politécnico de Bragança (IPB), Campus Santa Apolónia, 5300-253 Bragança, Portugal
e-mail: clvaz@ipb.pt

Â. P. Ferreira
e-mail: apf@ipb.pt

C. B. Vaz
Centre for Management and Industrial Engineering (CEGI/INESC TEC), Porto, Portugal

Â. P. Ferreira
CISE-Electromechatronic Systems Research Centre, University of Beira Interior, Covilhã, Portugal

J. P. Almeida et al. (eds.), *Operational Research*, Springer Proceedings in Mathematics & Statistics 411,
https://doi.org/10.1007/978-3-031-20788-4_14

1 Introduction

The European green deal embodies an ambitious plan to fight climate change, aiming to achieve carbon neutrality by 2050. In this context, the European countries are under political and environmental pressure to secure energy supplies, reduce dependence on fossil fuels and improve energy efficiency. Transportation is the remaining sector where emissions are still above the 1990 emissions level, being the road transportation the highest emitter and the main share of transport energy demand [1]. The main elements of the strategic plan are the increasing efficiency of the transport system and the use of low-emission energy sources for transport, which can be accomplished by the deployment of electric mobility.

In the transport sector, the effectiveness of the actions that have been carried out is still below the target, although technological advances and societal changes towards electric mobility have been significant. Despite the global increasing share of electrically-chargeable vehicles (plug-in electric vehicles (PEV)), i.e., battery electric vehicles (BEV) or plug-in hybrid electric vehicles (PHEV), the EU market share is still small [2]. The deployment of electric mobility is largely affected by regional and national-level policies, such as penetration of renewable electricity and charging infrastructure, among others [1].

A literature survey previously done on the assessment of the deployment of electric mobility [1] identified several approaches and indicators, which have supported the development of this work. Regarding the methodological approaches, they vary considerably with the data set under analysis and objectives, being Data Envelopment Analysis (DEA) employed by Onat et al. [3] and Neves et al. [4]. The first work proposed a multi-criteria decision-support framework using the input-based DEA multiplier model to assess the BEV efficiency for each country state using multiple inputs (GHG emission, energy and water consumption and operation cost) and output data (the service provided per vehicle-miles traveled) and an agent-based modeling to estimate the future market share. Neves et al. [4] used the output-based DEA multiplier model, considering multiple inputs (capital, labor force, Industrial Production Index, crude oil price, electricity intensity, number of different BEV models in top 10 of new registered BEV in a country) and outputs (BEV market share and accumulated number of public policies supporting electric mobility) in a first stage analysis to calculate the efficiency scores of several European countries addressing electric mobility, followed by a fractional regression model to identify the significant determinants of electric mobility.

Decision makers would benefit from the development of tools able to indicate and support the best strategies to deploy electric mobility. Under this context, a composite indicator able to measure the performance of European countries in terms of road transport sustainability would support new strategies towards the existing targets and goals (Europe 2030, e.g.). This paper aims to assess European countries' performance in deploying zero and low-emission vehicles in road transportation using a composite indicator (CI) model based on DEA. The CI only looks at achievements (sub-indicators) of the countries without taking into account the inputs used [5]. This is a major contribution of the present study because, to the best of the authors' knowledge, this is the first study to use this approach to evaluate the performance of the electric mobility deployment in European Countries. The selection of the sub-indicators should gather sustainable aspects of road transport and the main characteristics of the assessed European countries. The derivation of the CI for each country under evaluation is performed through the Benefit of Doubt (BoD) model, proposed in [6]. Therefore, the CI aggregates the selected sub-indicators observed for each country in 2019, in which the weights of those sub-indicators are calculated endogenously through the BoD model. This model handles directly the sub-indicators characterized by an isotonic behaviour, where higher scores indicate better performance, and also the reverse sub-indicators, characterized by an anti-isotonic behaviour, i.e, those whose lower scores translate into better performance.

It should be pointed out the fact that a DEA model endogenously determines the most favourable weights to maximize the performance for each country, for a given set of observations, which explains the appeal of DEA models to derive CI in real policy-related settings [7]. Additionally, to ensure the sustainability of the road mobility, European policy issues have to balance the general environmental concerns and the country-specific policy priorities. Therefore, the assessment of road mobility performance requires the comparison of the multi-dimensional feature of road mobility in EU members. To do so, the use of a fixed set of weights is not considered a fair judgement taking into account the uniqueness of each country. This issue is also framed in [8], as follows "in the context of the EU, there are evident difficulties in reaching agreement on such weights, given that each member state has its own national specificity".

The outline of this paper is as follows. Section 2 presents the performance assessment methodology and the data set adopted to calculate the composite indicators. Section 3 presents the results and the discussion of the findings. Finally, Sect. 4 closes the paper with the main conclusions and suggestions for future research.

2 Methods and Data

The following subsection provides an overview on the methodological approach to perform the assessment of European countries on the deployment of low-emission vehicles towards sustainable mobility. It describes BoD model proposed by [6] to determine the CI through the DEA methodology and the initial BoD model proposed by [5]. Afterwards, the Sect. 2.2 describes the set of relevant sub-indicators which are selected to assure a fair performance comparison among countries.

2.1 Methods

Cherchye et al. [5] initially proposed the BoD model to handle isotonic sub-indicators. This model is equivalent to the original input-based DEA multiplier model [9], since all sub-indicators are considered as outputs and a single dummy input equal to one is considered for all units.

Considering a cross-section of m isotonic sub-indicators i for each unit $j (j = 1, \ldots, s)$, being y_{ij} the score of that sub-indicator observed for each unit j, and w_i the weight assigned to it. The BoD model (1) enables to evaluate the performance for each unit under assessment j_0, by computing the composite indicator (CI_{jo}) through the weighted average of m sub-indicators, in which the optimum weights are endogenously calculated.

$$\begin{aligned} \mathrm{CI}_{j_0} &= max \sum_{i=1}^{m} w_i y_{ij_0} \\ \textbf{\textit{s.t.}} \quad & \sum_{i=1}^{m} w_i y_{ij} \leq 1 \quad \forall j = 1, ..., s \\ & w_i \geq 0 \quad \forall i = 1, ..., m \end{aligned} \tag{1}$$

Thus, the model (1) determines the most advantageous w_i for each sub-indicator y_{ij_0} that maximizes the CI_{j_0} score for the country j_0 by comparison with the best practices frontier. Therefore, each country cannot claim that a poor relative performance is due to a harmful or unfair weighting scheme, as the model (1) determines the optimum w_i which maximizes its CI_{j_0} by comparison with best practices frontier, as the w_i is endogenously estimated by this model [5].

To avoid the situations in which high performance scores are achieved, mainly due to assigning zero weights to some sub-indicators (for example in case of the sub-indicator has a relative poor performance), owning no influence in the road mobility performance or emphasizing weights, proportional virtual weight restrictions (2) are imposed to the unit under evaluation [10]. In the present context, the proportional virtual weight restrictions are particularly interesting as these are independent on measurement units and directly show how the respective pie shares contribute to a CI_{j_0} score [7].

$$\alpha \leq \frac{w_i y_{ij_0}}{\sum_{i=1}^{m} w_i y_{ij_0}} \leq \beta \quad \forall i = 1, ..., m \tag{2}$$

Restriction (2) imposes that each sub-indicator is required to have minimum and maximum percentages of contribution, α and β, respectively, in the assessed composite indicator for the unit under evaluation. This type of restriction ensures that the virtual weights of the DMU under evaluation (j_o) are within the set limits, although the virtual weights for the other DMUs j may not be within the limits. Consequently, some inefficient DMUs ($\mathrm{CI} < 1$) when evaluated using the specific system of weights of the DMU under evaluation may appear as its peers. Despite this, it is one of the most used approaches to restrict virtual weights [6, 7], since restricting the weights of all DMUs would lead to infeasibility issues [11, 12].

It is frequent that the performance assessment has to handle anti-isotonic sub-indicators, as for example road GHG emissions per capita, in the road mobility assessment. To use the BoD model (1), it is required to perform a mathematical transformation on these anti-isotonic sub-indicators before incorporating them in the model. This approach has some drawbacks as discussed in [11], being preferable to avoid the data transformation by using models that handle directly the anti-isotonic sub-indicators. For example, Fare et al. [6] consider the anti-isotonic sub-indicators as a reverse. Other works, such as [13–17] apply the directional distance function model to increase the isotonic sub-indicators and decrease the anti-isotonic sub-indicators, in a similar way. The present study adopts the approach proposed by Färe, Karagiannis, Hasannasab and Margaritis [6] (hereinafter named FKHM model) to assess the performance of road mobility of European countries.

The FKHM model handles the anti-isotonic sub-indicators as reverse rather than as undesirable. According to Fare et al. [6], the presence of isotonic sub-indicators does not imply nor is implied by the presence of reverse sub-indicators, being the variation of reverse sub-indicators independent on the values of isotonic sub-indicators. Additionally, a cross-section of reverse sub-indicators y_{ij} $(i = m + 1, \ldots M)$ is considered concerning each country j. Thus, the FKHM model is formulated as (3), where y_{ij} $(i = 1, ..., m)$ are the isotonic sub-indicators (i.e., capturing positive aspects) while y_{ij} $(i = m + 1, \ldots M)$ are the reverse sub-indicators (i.e., capturing negative aspects).

$$
\begin{aligned}
\mathrm{CI}_{j_0} &= max \sum_{i=1}^{m} w_i y_{ij_0} - \sum_{i=m+1}^{M} w_i y_{ij_0} \\
\boldsymbol{s.t.} \quad & \sum_{i=1}^{m} w_i y_{ij} - \sum_{i=m+1}^{M} w_i y_{ij} \leq 1 \quad \forall j = 1, ..., s \\
w_i &\geq 0 \quad \forall i = 1, ..., M
\end{aligned}
\tag{3}
$$

The FKHM model determines the optimum weights for the country under assessment by maximizing its CI_{j_0}, which is the difference between the weighted average of isotonic sub-indicators and the weighted average of reverse sub-indicators, since that difference has a maximum of 1 for each DMU observed. Thus, for each DMU under assessment (j_o), all DMU j are evaluated with the same weight w_i for each isotonic and reverse sub-indicator y_{ij}. It should be pointed out that the FKHM model can be simplified to the BoD model if no anti-isotonic indicators are considered [6].

The use of weights restrictions should combine some degree of flexibility (unrestricted DEA is the most flexible alternative) and the suitable consistency that accomplishes that all dimensions are taken into account. Following this idea, some importance should be given to the combined effect between flexibility and consistency [18] which is controlled by the k score. Taken into account that in this study, no expert information is available, and considering M sub-indicators, constraint (2) should guarantee that the proportional virtual weight for each sub-indicator should vary between $\frac{1}{M}(1-k)$ and $\frac{1}{M}(1+k)$, with $k \in]0,1[$. The selection of k is performed through a robustness analysis in the model results. Thus, the derived model (4) is used to assess the performance of road mobility in European countries.

$$
\begin{aligned}
\mathrm{CI}_{j_0} &= max \sum_{i=1}^{m} w_i y_{ij_0} - \sum_{i=m+1}^{M} w_i y_{ij_0} \\
\boldsymbol{s.t.} \quad & \sum_{i=1}^{m} w_i y_{ij} - \sum_{i=m+1}^{M} w_i y_{ij} \leq 1 \qquad \forall j = 1, ..., s \\
\frac{1}{M}(1-k) &\leq \frac{w_i y_{ij_0}}{\sum_{i=1}^{M} w_i y_{ij_0}} \leq \frac{1}{M}(1+k) \qquad \forall i = 1, ...M, k \in]0, 1[\\
w_i &\geq 0 \qquad \forall i = 1, ..., M
\end{aligned} \tag{4}
$$

2.2 Data

The performance assessment considers data from the year 2019, before the COVID-19 pandemics. In fact, data and estimates for 2020 indicate a significant activity decrease, which would bias the analysis. The countries under analysis are the EU Member States in 2019, i.e., the actual 27 EU Member States plus the United Kingdom. To assess the performance of EU-28 with regard to the deployment of low-emission vehicles, the sub-indicators should incorporate the impacts, trends and technological advances in mobility. In addition, their selection should favour a fair comparison between units, while translating the specificities of each country. Keeping this in mind, the selection of the sub-indicators was performed exploiting the existing literature, the available data for the European countries under study and the requirement of measuring a single aspect. Seven sub-indicators ($M = 7$) are selected, five isotonic and two reverse, described below. These sub-indicators must reflect the outputs achieved normalized to assure a fair comparison between countries. Table 1 summarizes these sub-indicators, their measurement units and the sources.

The market share of plug-in electric vehicles (PEV) (*PEV Market Share*) is given by the percentage of newly registered battery electric vehicles (BEV) and plug-in hybrid electric vehicles (PHEV) relative to the total newly registered passenger vehicles (M1 category) in the year under analysis. Most European countries have their sales market share still dominated by petrol and diesel vehicles. The penetration of vehicles powered by one or more electric motors reflects the policies and societal changes, which is essential to assess and translate the deployment of the electric mobility, considering, for instance, the increasing supply of electric propulsion models, tax policies, environmental consciousness, among others.

Table 1 Summary of the sub-indicators

Sub-indicators	Description	Unit	Source
Isotonic			
PEV Market Share	Market share of new registrations of PEV	%	EAFO
Renewable Energy	Share of energy from renewable sources in transport	%	EUROSTAT
Public Transport	Share of buses and trains in total passengers' transport	%	EUROSTAT
GDP per capita	GDP per capita (chained linked volumes 2010)	€ per capita	EUROSTAT
Railway Length	Total length of railway lines per country's area	km^{-1}	EUROSTAT
Reverse			
GHG Emissions	GHG emissions from fuel combustion in road transport	Tonnes/inhabitant	EUROSTAT
New Car Emissions	Average CO_2 emissions from new passengers' cars	gCO_2/km	EUROSTAT

EAFO [19], EUROSTAT [20]

The share of energy from renewable sources in transport (*Renewable Energy*) specifies the percentage of renewable energy in the total transport fuels. The European SHARES tool manual [21] provides a guide for harmonised calculation of the share of energy from renewable sources, being the renewable energy in transportation defined by sustainable biofuels, renewable electricity, hydrogen and synthetic fuels of renewable origin such as geothermal, solar thermal, renewable municipal waste and solid biofuels. A higher share of renewable energy favours the energy independency in the transport sector while contributing to a significant reduction in GHG emissions, and also reducing the local air and noise pollution. The European Transport Roadmap [22] suggests a regular phase out of conventionally-fuelled vehicles from urban environments while the Renewable Energy Directive from 2018 [23] sets a 32% target for the penetration of renewable energy in the transport sector, by 2030.

The share of buses and trains in total passengers' transport (*Public Transport*) is defined by the percentage of passenger's transport that is made by buses (including coaches and trolley-buses) and trains relative to the total inland transport (passenger cars, buses and trains). Trams and metros are not included due to the lack of harmonised data. The passengers' transport data is measured in passenger-kilometer (pkm), a unit that represents the transport of one passenger over one kilometer using a specific mode of transport.

The Gross Domestic Product per capita (*GDP per capita*), calculated as the ratio of the GDP to the average population in the 2019 year, provides a measure of the

Table 2 Descriptive statistics for the selected sub-indicators

	PEV Market Share	*Renewable Energy*	*Public Transport*	*GDP per capita*	*Railway Length*	*GHG Emissions*	*New Car Emissions*
Mean	0.029	0.088	17.757	27840.357	0.053	2.101	122.521
Std.Dev.	0.034	0.053	4.277	17246.124	0.032	1.606	9.474
Max	0.149	0.303	28.400	83640.000	0.121	10.031	137.600
Min	0.004	0.033	9.400	6840.000	0.013	0.935	98.400

economic activity through the total final output of goods and services produced within the year under analysis. High values of this sub-indicator reflect high levels of economic productivity of the economies of the countries able, for instance, to support increased salaries and tax policies boosting the electric mobility.

The total length of railway lines indicator (*Railway Length*) gives the total length of railways, electrified or not, on the territory of the respective country. The original data is given in kilometers and is normalized using each country total area in km^2, to account for its dimension. This indicator reflects an infrastructure able to provide a more sustainable transport, due to high-volume transport feature.

The GHG emissions from fuel combustion in road transport (*GHG Emissions*) indicator measures the contribution of the road transport emissions to the total GHG emissions inventory. The GHG emissions include those of carbon dioxide (CO_2), methane (CH_4), nitrous oxide (N_2O), perfluorocarbons (PFCs), hydrofluorocarbons (HFCs), sulphur hexafluoride (SF_6) and nitrogen trifluoride (NF_3). The values, originally expressed in thousand tonnes, are normalized using the countries' population on the 1st January of the year under analysis, to take into consideration their dimension.

The average carbon dioxide (CO_2) emissions from new passengers' cars (*New Car Emissions*) is expressed in grams of CO_2 per kilometer. The Regulation (EU) 2019/631 set a mandatory target for emissions reduction of 95 g CO_2/km for new passenger cars, by 2021 [24]. This target applies for the average of the manufacturer's overall fleet, i.e., cars above this limit are allowed in the market as long as they are offset by the production of cars with lower emissions. This sub-indicator is independent of the previous one, complementing it by reflecting the emissions levels of new sales.

Table 2 presents the descriptive statistics of the sub-indicators in the year under analysis (2019).

3 Results and Discussion

This section presents the results concerning the performance assessment of the electric mobility deployment in EU-28 countries, by using the model (4) and the selected sub-indicators ($M = 7$) observed in 2019, previously described.

Initially, some different weight schemes were considered in the model (4), by ranging the score of k to explore the robustness of performance results. Then, the more robust scenario was adopted to assess the performance of road mobility in European countries. Additionally, the relative strengths and weaknesses of each country in terms of road sustainability are further explored.

3.1 *Robust Proportional Virtual Weight Restrictions*

To explore the robustness of performance scores, some different scenarios were considered in the model (4), setting k equal to 0.60, 0.65, 0.70, 0.75, 0.80, 0.85, 0.90 and 0.95. Scores lower than 0.6 were not considered given that only two countries were assessed as efficient (Netherlands and Sweden) for $k = 0.5$. Since those weight limits are arbitrary, the robustness of the performance scores was analysed by modifying the weight constraints between different limits. The limit values of the considered range were not used because the first one, $k = 0$, implies no flexibility and $k = 1$ implies that the lower limit of the weight for each sub-indicator may become 0, which would result in countries disregarding some sub-indicators. Table 3 presents the CI results for the eight weighting schemes for each country which allows to understand the robustness of the CI results. Figure 1 shows the box plot of CI results for each country, being the mean of the eight scenarios plotted by the black square.

Globally, it is observed that the CI results are robust for the majority of underperforming countries (i.e., never achieve CI = 1 with any k), since slight variations were detected except in $k = 0.95$ scenario, where higher variations were observed. In fact, this scenario allows the largest level of flexibility in weights calculation, from which two countries become efficient (Ireland and Hungary). Ireland was assessed with CI = 0.942 in the $k = 0.90$ scenario, while Hungary was assessed with CI = 0.888. Noteworthy, is the fact that Ireland is only used once as benchmark while Hungary is not used as benchmark by any other country in $k = 0.95$ scenario. Thus, those countries were significantly affected by the higher level of flexibility such as the underperforming countries Czech Republic and Slovakia as it is shown in Fig. 1.

From the above, a decision was made to consider the $k = 0.90$ scenario in further analyses, aiming to achieve a trade-off between robustness scores, flexibility and consistency on weights calculation. Using this scenario, the model (4) assures that the proportional virtual weight for each sub-indicator should vary between 1.43% and 27.14%.

The dual of the model (4) was used to identify the benchmarks for underperforming countries. Within this scenario, four benchmarks (i.e., CI = 1) were identified: Belgium, Luxembourg, Netherlands and Sweden. The CI average equals to 0.638 with a standard deviation of 0.246, which indicate that several EU-28 countries have potential to improve their practices towards better road transport sustainability. A grayscale was applied to classify the countries in terms of worst, medium and best road mobility performance, respectively, as shown in Fig. 2.

Table 3 CI results for $k = 0.60, 0.65, 0.70, 0.75, 0.80, 0.85, 0.90$ and 0.95 in model (4)

	CI							
Country	$k = 0.60$	$k = 0.65$	$k = 0.70$	$k = 0.75$	$k = 0.80$	$k = 0.85$	$k = 0.90$	$k = 0.95$
Belgium	0.732	0.783	0.828	0.872	0.918	0.968	1.000	1.000
Bulgaria	0.265	0.283	0.304	0.327	0.352	0.380	0.412	0.449
Czechia	0.299	0.330	0.366	0.405	0.447	0.501	0.574	0.835
Denmark	0.678	0.697	0.711	0.726	0.747	0.778	0.845	0.886
Germany	0.686	0.724	0.765	0.806	0.848	0.887	0.909	0.944
Estonia	0.203	0.223	0.244	0.269	0.297	0.326	0.350	0.383
Ireland	0.631	0.647	0.666	0.695	0.733	0.820	0.942	1.000
Greece	0.208	0.224	0.241	0.260	0.279	0.295	0.315	0.340
Spain	0.411	0.432	0.453	0.467	0.484	0.502	0.520	0.533
France	0.616	0.639	0.660	0.678	0.697	0.708	0.714	0.722
Croatia	0.263	0.282	0.304	0.324	0.344	0.368	0.399	0.439
Italy	0.409	0.442	0.478	0.517	0.559	0.599	0.648	0.700
Cyprus	0.260	0.267	0.274	0.282	0.291	0.299	0.307	0.320
Latvia	0.257	0.274	0.292	0.307	0.321	0.339	0.360	0.379
Lithuania	0.209	0.225	0.244	0.262	0.273	0.286	0.301	0.316
Luxembourg	0.876	0.933	0.995	1.000	1.000	1.000	1.000	1.000
Hungary	0.450	0.473	0.520	0.587	0.679	0.817	0.880	1.000
Malta	0.403	0.413	0.419	0.426	0.432	0.439	0.446	0.452
Netherlands	1.000	1.000	1.000	1.000	1.000	1.000	1.000	1.000
Austria	0.749	0.779	0.805	0.830	0.853	0.874	0.888	0.908
Poland	0.254	0.276	0.301	0.327	0.354	0.384	0.423	0.482
Portugal	0.479	0.485	0.491	0.496	0.501	0.506	0.511	0.515
Romania	0.363	0.389	0.422	0.466	0.523	0.583	0.611	0.634
Slovenia	0.370	0.397	0.425	0.451	0.472	0.497	0.528	0.558
Slovakia	0.249	0.278	0.313	0.348	0.390	0.438	0.494	0.675
Finland	0.733	0.741	0.749	0.755	0.760	0.766	0.773	0.779
Sweden	1.000	1.000	1.000	1.000	1.000	1.000	1.000	1.000
United Kingdom	0.629	0.653	0.676	0.697	0.706	0.713	0.719	0.725
Mean	0.489	0.510	0.534	0.556	0.581	0.610	0.638	0.678
Std.Dev.	0.246	0.247	0.248	0.245	0.243	0.245	0.246	0.247
Minimum	0.203	0.223	0.241	0.260	0.273	0.286	0.301	0.316
No. of efficient units	2	2	2	3	3	3	4	6

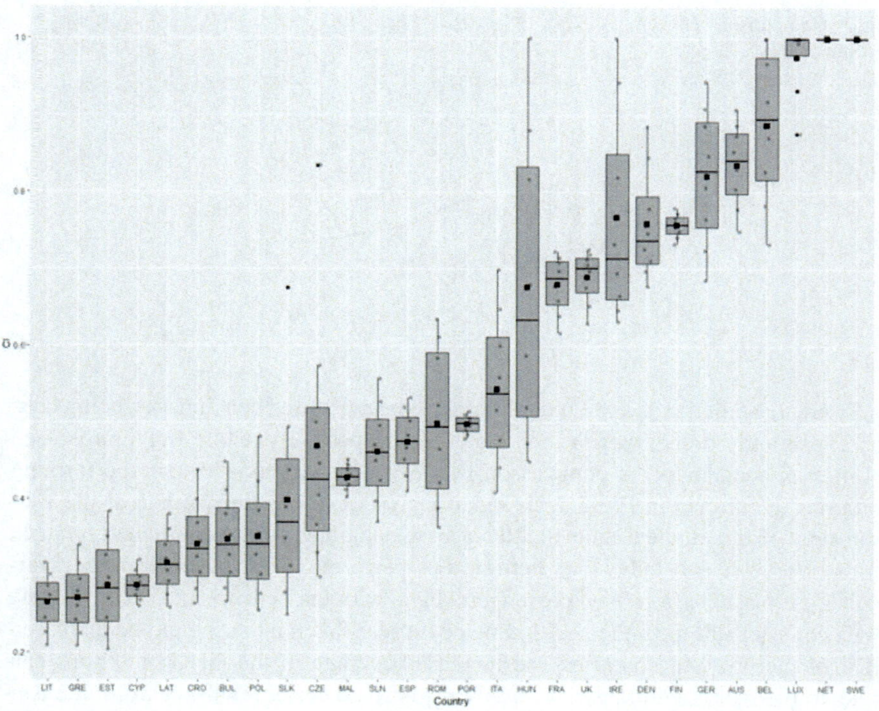

Fig. 1 Robustness analysis to k changes ($k = 0.60, 0.65, 0.70, 0.75, 0.80, 0.85, 0.90$ and 0.95) in model (4)

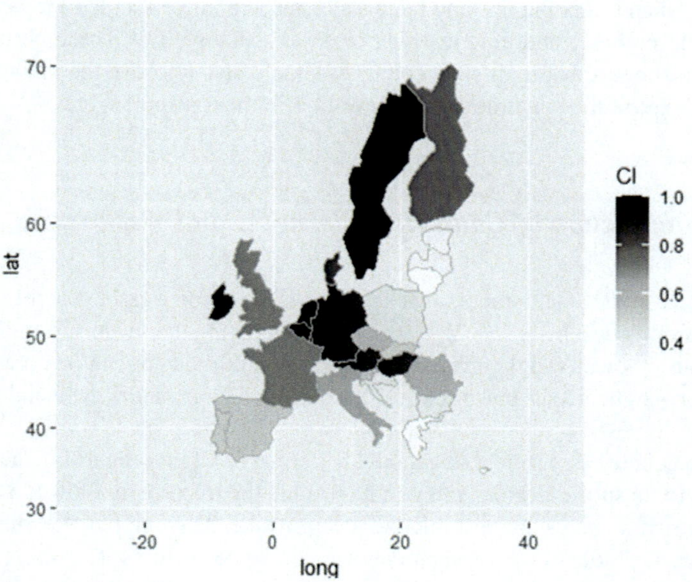

Fig. 2 CI results (model (4) with $k = 0.90$)

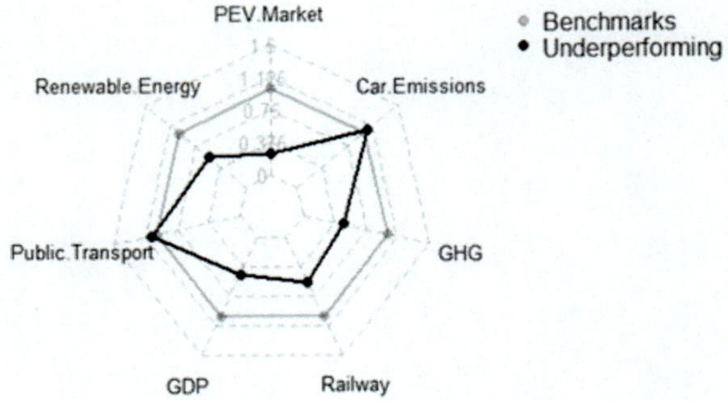

Fig. 3 Comparison between benchmarks and underperforming countries

Sweden, Netherlands and Luxembourg are the most used benchmarks, being used for 23, 18 and 16 underperforming countries, respectively, while Belgium is used 5 times as benchmark. A comparison between benchmarks and underperforming countries is outlined in Fig. 3. This radar analysis exhibits the average of the sub-indicators of the efficient units and the average obtained for the inefficient ones. In terms of reverse sub-indicators, both groups perform similar in terms of the average CO_2 emissions from new passengers' cars, with the benchmarks exceeding the GHG emissions of the underperforming countries. Taken into account the increased GDP of the efficient countries, the increased average of the GHG emissions can relate to the increased industrial activity of those countries. With regard to the isotonic sub-indicators, underperforming countries obtain the worse result compared with benchmarks in the market share of electrically chargeable vehicles, suggesting these countries would benefit from policies and technological advances (for instance, incentives, such as tax breaks, and increased charging infrastructures) to better support electric mobility and, in consequence, road sustainability. These performance results also suggest that inefficient countries should also improve the railway sector and better exploit the penetration of renewables in the transport sector.

3.2 Identification of Countries' Strengths and Weaknesses

The adopted model (4) compels that the sum of share of weight assigned to each sub-indicator i given by $w_i y_{ij_0} / \sum_{i=1}^{M} w_i y_{ij_0}$, considering the seven sub-indicators is 100%. Thus, a lower weight attributed to a given sub-indicator can be compensated by a higher weight at least one or more other sub-indicators, under the weights' range imposed.

In each country assessment, the model has some flexibility regarding the weights attributed to its sub-indicators and could attribute the maximum allowed weight on them in which it shows the best relative performance. This implies that the optimal weight structure identified for each country can be relevant as it is affected by its relative strengths and weaknesses. Figure 4 presents the optimal weights' structure

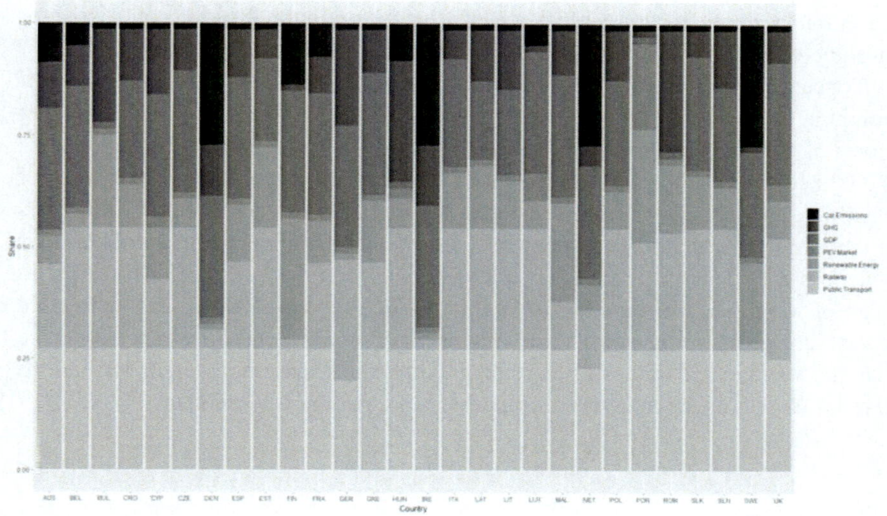

Fig. 4 Optimal weight of share structures for $k = 0.90$ in weight constraint (4)

given by the share of weight $w_i y_{ij_0} / \sum_{i=1}^{M} w_i y_{ij_0}$ assigned to sub-indicator i for each country j_o, which enables to identify the different countries profiles in terms of their relative strengths and weaknesses. In fact, the country profile cannot be unique if there are alternative optimum solutions, as the optimal weight structure obtained is one of the possible solutions [25]. This issue tends to affect more the benchmarks than the underperforming countries.

To maximize the road mobility performance, the model allocates, on average, a share of weights equal to 83% for the isotonic sub-indicators while that for the reverse indicators is 17%.

Regarding the isotonic sub-indicators, the maximum share of weight is assigned to *Public Transport* for almost all countries, allowing the higher consistency among countries that put the maximum weight in this sub-indicator. Similar maximum share of weight occurs with the *GDP per capita*, with exception of Bulgaria, Hungary, Portugal and Romania, which put the minimum weight in this sub-indicator. The third sub-indicator more weighted on underperforming countries is the *Railway Lenght*, except in Denmark, Ireland and Finland, which have lower scores than the average. The fourth isotonic sub-indicator more weighted is the *Renewable Energy*. Finland, Portugal and Malta are the underperforming countries that are evaluated with the highest share of weight (>22%) in this sub-indicator. In fact, Finland and Portugal are in the top 6 position in *Renewable Energy* and Malta is on the average position. Globally, the minimum share of weight is attributed to *PEV Market Share* with exception of Portugal which achieves the maximum share of weight equal to 19.1%. All underperforming countries put the minimum share of weight on the electric mobility, except Cyprus, Hungary and UK, in which it ranges between 2.6% and

6.2%. Furthermore, it is possible that an alternative optimum solution could exist for Sweden to highlight its second higher score on the *PEV Market Share*.

Concerning the reverse sub-indicators, Portugal assigns the minimum share of weight at each of them (total share of weight is 2.86%). In fact, Portugal has lower scores of *GHG Emissions* and *New Car Emissions* than the average. Since Portugal is on the top 10 position regarding electric vehicle deployment and in penetration of renewable energy in the transport sector, the contribution of the related sub-indicators in the CI is 19% and 25%, respectively. Additionally, the contribution of *Railway Length* is 24% where Portugal has a lower score than the average. The maximum share of weights is assigned to *Public Transport*, although Portugal has the second lowest score. This weight' structure explains why Portugal owns a lower performance than the average (CI = 0.511). Further research should explore weight restrictions in order to give different levels of importance among the sub-indicators.

4 Conclusions

The obtained CI results demonstrated to be robust for variations of k, considering the majority of inefficient countries except in the scenario $k = 0.95$, where higher variations were observed. In order to consider a trade-off between robustness scores, flexibility and consistency on weights calculation, the $k = 0.90$ scenario was adopted in the methodological approach performed.

Four benchmarks (CI = 1) were identified (Belgium, Luxembourg, Netherlands and Sweden). It was observed an average of CI equal to 0.638 with a standard deviation of 0.246 which indicate that several European countries have potential to improve their practices towards better road transport sustainability.

Most countries have potential to improve road transport sustainability for instance, by following the best practices adopted by the benchmarks. The performed analysis highlights that inefficient countries should take measures to better support the penetration of PEV vehicles (for instance, incentives, such as tax breaks, and increased charging infrastructures). In addition, the obtained results also suggest that the railway sector and the penetration of renewable energies are the remaining areas in the transport sector that should be enhanced.

The model (4) allocates, on average, a share of weights equal to 83% for the isotonic sub-indicators while the share for the reverse indicators is 17%, in which the lowest average share is attributed to the electric vehicle deployment (2%).

The optimal weights' structure identified for each country could be relevant as it is affected by its relative strengths and weaknesses. These results show the flexibility of the model (4) in choosing the best range of weights' structure to maximize the relative road mobility performance of each country, translating its different specificities. Note that this procedure can be affected by the possible alternative optimum solutions, when at least another optimal structure exists.

Further research should explore the dynamic performance to identify trends by considering the data over the years, and also other European countries that are restricted to the same transport policy. Additionally, the determinants of the performance should be explored to identify factors capable of influencing policy making. Since the adopted model considers the same lower and upper bounds for the proportion of weights attributed to all sub-indicators, some different weights schemes can be defined in order to consider different bounds, enabling to rank the sub-indicators importance.

References

1. Gruetzmacher, S.B., Vaz, C.B., Ferreira, Â.P.: Assessing the deployment of electric mobility: a review. In: Gervasi, O., et al. (eds.) Computational Science and Its Applications—ICCSA 2021. Lecture Notes in Computer Science, vol. 12953, pp. 350–365. Springer, Cham (2021)
2. IEA: Global ev outlook 2018. Technical report (2018). www.iea.org/reports/global-ev-outlook-2018
3. Onat, N.C., Noori, M., Kucukvar, M., Zhao, Y., Tatari, O., Chester, M.: Exploring the suitability of electric vehicles in the United States. Energy **121**, 631–642 (2017)
4. Almeida Neves, S., Cardoso Marques, A., Moutinho, V.: Two-stage DEA model to evaluate technical efficiency on deployment of battery electric vehicles in the EU countries. Transp. Res. Part D: Transp. Environ. **86**(August), 102–489 (2020)
5. Cherchye, L., Moesen, W., Rogge, N., Van Puyenbroeck, T.: An introduction to 'benefit of the doubt' composite indicators. Soc. Indic. Res. **82**(1), 111–145 (2007)
6. Färe, R., Karagiannis, G., Hasannasab, M., Margaritis, D.: A benefit-of-the-doubt model with reverse indicators. Eur. J. Oper. Res. **278**(2), 394–400 (2019)
7. Cherchye, L., Moesen, W., Rogge, N., Van Puyenbroeck, T., Saisana, M., Saltelli, A., Liska, R., Tarantola, S.: Creating composite indicators with dea and robustness analysis: the case of the technology achievement index. J. Oper. Res. Soc. **59**(2), 239–251 (2008)
8. Atkinson, A., Cantillon, B., Marlier, E., Nolan, B.: Social indicators. The EU and Social Inclusion. Oxford University Press, Oxford (2002)
9. Charnes, A., Cooper, W.W., Rhodes, E.: Measuring the efficiency of decision making units. Eur. J. Oper. Res. **2**(6), 429–444 (1978)
10. Wong, Y.H., Beasley, J.: Restricting weight flexibility in data envelopment analysis. J. Oper. Res. Soc. **41**(9), 829–835 (1990)
11. Dyson, R.G., Allen, R., Camanho, A.S., Podinovski, V.V., Sarrico, C.S., Shale, E.A.: Pitfalls and protocols in dea. Eur. J. Oper. Res. **132**(2), 245–259 (2001)
12. Sarrico, C., Dyson, R.: Restricting virtual weights in data envelopment analysis. Eur. J. Oper. Res. **159**(1), 17–34 (2004)
13. Charles, V., Färe, R., Grosskopf, S.: A translation invariant pure dea model. Eur. J. Oper. Res. **249**(1), 390–392 (2016)
14. Fusco, E.: Enhancing non-compensatory composite indicators: a directional proposal. Eur. J. Oper. Res. **242**(2), 620–630 (2015)
15. Vidoli, F., Fusco, E., Mazziotta, C.: Non-compensability in Composite Indicators: a Robust Directional Frontier Method. Soc. Indic. Res. **122**, 635–652 (2015)
16. Zanella, A., Camanho, A.S., Dias, T.G., Camanho, A.S.: The assessment of cities' livability integrating human wellbeing and environmental impact. Ann. Oper. Res. **226**, 695–726 (2015)
17. Zanella, A., Camanho, A.S., Dias, T.G.: Undesirable outputs and weighting schemes in composite indicators based on data envelopment analysis. Eur. J. Oper. Res. **245**(2), 517–530 (2015)
18. Cárcaba, A., González, E., Ventura, J.: Social progress in Spanish municipalities (2001–2011). Appl. Res. Qual. Life **12**, 997–1019 (2017)
19. European Alternative Fuels Observatory. https://www.eafo.eu/. Accessed 20 Sept 2021

20. European statistical office. https://ec.europa.eu/eurostat/web/main/data/database. Accessed 28 Sept 2021
21. European Commission: SHARES tool manual. Unit E.5: Energy (2018)
22. European Comission: White paper on transport: roadmap to a single European transport area: towards a competitive and resource-efficient transport system (2011)
23. European Commission: Directive 2018/2001 of the European parliament and of the council. Official J. Eur. Union (2018)
24. Commission, E.: Regulation (EU) 2019/631 of the european parliament and of the council. Official J. Eur. Union (2019)
25. Amado, C.A., São José, J.M., Santos, S.P.: Measuring active ageing: a data envelopment analysis approach. Eur. J. Oper. Res. **255**(1), 207–223 (2016)

The Art of the Deal: Machine Learning Based Trade Promotion Evaluation

David Branco Viana and Beatriz Brito Oliveira

Abstract Trade promotions are complex marketing agreements between a retailer and a manufacturer aiming to drive up sales. The retailer proposes numerous sales promotions that the manufacturer partially supports through discounts and deductions. In the Portuguese consumer packaged goods (CPG) sector, the proportion of price-promoted sales to regular-priced sales has increased significantly, making proper promotional planning crucial in ensuring manufacturer margins. In this context, a decision support system was developed to aid in the promotional planning process of two key product categories of a Portuguese CPG manufacturer. This system allows the manufacturer's commercial team to plan and simulate promotional scenarios to better evaluate a proposed trade promotion and negotiate its terms. The simulation is powered by multiple gradient boosting machine models that estimate sales for a given promotion based solely on the scarce data available to the manufacturer.

Keywords Promotion planning · Retail · Trade promotion

1 Introduction

Over the last half-decade, the intensity of price discounts given as part of trade promotions between retailers and manufacturers, as well as the share of total sales these promotions represent, has been rapidly growing. The four largest British supermarkets reported that, in 2016, in the consumer packaged goods (CPG) sector, 45% of the average consumer's expenditure was on price-promoted products [11]. Nielson

D. B. Viana (✉) · B. B. Oliveira
Faculty of Engineering of the University of Porto, Porto, Portugal
e-mail: david.viana@ltplabs.com

D. B. Viana
LTPLabs, Senhora da Hora, Portugal

B. B. Oliveira
INESC TEC, Porto, Portugal

J. P. Almeida et al. (eds.), *Operational Research*, Springer Proceedings in Mathematics & Statistics 411,
https://doi.org/10.1007/978-3-031-20788-4_15

[13] found that, in 2018, promotional sales accounted for half of the Portuguese CPG sector's sales. For comparison, promotional sales accounted only for a sixth of total sales for the Spanish counterpart [13]. With retailers proposing such a significant amount of trade promotions to manufacturers, assessing how profitable or worthwhile a given promotion will be becomes paramount.

In this context, a CPG manufacturer is interested in developing a custom-made decision support system (DSS) to aid in its promotional planning process for its two main product categories: olive oil and vegetable oil. The main objective of this work is to present the DSS developed. This includes the forecasting models trained to accurately predict promotional sales for the various retailers and categories, despite only leveraging the limited data available to the manufacturer. Considering this limitation and how significant promotional sales are in the total sales figure, this work also aims to assess whether it is possible to detect undesirable deals before they happen. The decision support system presented was developed in the context of a consulting firm, for a project with a Portuguese CPG manufacturer.

This paper is structured as follows. The remainder of this section describes the company, its retail partners, and its market, and introduces key concepts applied in the paper. Section 2 details the decision support system implemented, as well as the data used and the models trained on such data. Section 3 elaborates on the models obtained, their performance and metrics, and compares these results, drawing conclusions about the different retailers and categories and describing the final deployment of the system. Section 4 concludes the document, summarizing the purpose, methodology and contributions of this work.

1.1 The Manufacturer-Retailer Promotion Dynamics

The manufacturer considered in this work deals with numerous retailers in several countries and offers multiple product lines across various CPG categories. In this work, the Portuguese olive oil and vegetable oil markets were considered, in both of which the manufacturer is the market leader. The company supplies roughly 75% and 45% of the total vegetable oil and olive oil market demand (in liters), respectively. Roughly 40% of the vegetable oil market and 20% of the olive oil market is controlled by the manufacturer's brands, with the remaining 35% and 25%, respectively, being split among its store brands.[1] It is important to note that the olive oil market is more heavily disputed than the vegetable oil market, with more competitors and promotions, as will be discussed later.

This work focuses on two key retailers (retailer A and retailer B), each controlling approximately 20% of the Portuguese CPG market share. Retailer A provided the manufacturer with sell-out data, whereas for retailer B only sell-in data is available.

---

[1] Store brand products, also called own brand products, are produced by a manufacturer for resale under a brand controlled by a retailer.

Sell-in is defined as units the manufacturer sells to the retailers, whereas sell-out is defined as units the retailer sells to the customers.

Manufacturers and retailers make deals among themselves to offer lower prices for certain products to customers and increase their sales through trade promotions. Although trade promotions can take on several forms, for this paper's purposes, a trade promotion is a partnership between a retailer and a manufacturer involving a sales promotion directed at the customer, in the form of a discount, and one directed at the retailer, in the form of allowances or discounts. The degree to which the discount offered to the retailer is passed onto the customers is called pass-through.

$$\text{Pass-through} = \frac{\text{Customers' total savings due to the trade promotion}}{\text{Retailer's total savings due to the trade promotion}}$$

Additionally, the retailer may also take advantage of the discount and stockpile products (also known as forward buying), which is an important source of profit for the retailer [4]. However, this does not directly benefit the manufacturer since no incremental sales are generated despite the discount offered, negatively impacting the manufacturer's bottom line. Similarly, the customer is not benefited since there is ultimately no discount. A successful trade promotion is one where additional sales (also called incremental sales) are generated beyond the usual sales level (the baseline) such that both parties profit, even considering the costs incurred in the promotion (such as discounts or marketing expenses). Baseline sales are the sales which would have been recorded had the promotion not taken place. This success is also expressed as the ratio between the incremental sales and the baseline sales, called the sales lift, generally expressed as a percentage.

The Portuguese consumer packaged goods market is dominated by sales promotions. In 2018, promotional sales accounted for half of the sector's sales [13]. Due to the pressure from competing manufacturers and the retailers operating with them, there has been an increase in the promotional frequency and the promotional intensity in recent years. The former is defined as the percentage of time that discounts are offered by retailers while the latter is defined as the average depth of those discounts. Manufacturer data points to even higher promotional frequency values in the categories of interest, as can be seen in Table 1, and also to an upward trend in promotional intensity, as shown in Table 2. These increases have lead to growing promotional saturation in the market, that is, the novelty factor of a given sales promotion is being reduced. This increase in promotional saturation also incentivizes customers to stockpile and wait for future promotions to resupply.

Regarding the company's promotional planning process, at the end of every year, as part of its annual corporate goal setting process, the company defines the annual volume and margin targets and sets the Manufacturer's Suggested Retail Price (MSRP) for the various products in each product line. This target definition takes into consideration current and forecasted commodity prices for the year, for both olive oil and the various kinds of vegetable oil (e.g., sunflower, sesame). On a quarterly basis, the promotional planning team drafts a rough proposal for each one of the retailers, taking into consideration both the sales promotions held in the same

Table 1 Weighted average of percent promotional sales. The olive oil and vegetable oil figures are based on company branded product sales data, in retailer A, while the other categories' figures are based on sales data of comparable products and retailers

Year	Olive Oil (%)	Veg. Oil (%)	Cookies	Cereal	Beer	Yogurts
2016	32.5	70.1	–	–	–	–
2017	69.0	72.6	–	–	–	–
2018	87.2	72.8	43.0%	47.0%	77.0%	54.0%
2019	94.7	76.0	46.0%	50.0%	78.0%	68.0%
2020[a]	91.5	93.3	49.0%	50.0%	80.0%	66.0%

[a]2020 figures pertain to the first half of 2020 only

Table 2 Weighted promotional intensity over the years, using company branded product sales data, for both olive oil and vegetable oil

Category	Retailer	2016	2017	2018	2019	2020[a]
Olive Oil	Retailer A	3.8%	11.4%	26.7%	33.2%	40.6%
	Retailer B	11.3%	15.9%	26.6%	38.0%	44.7%
Vegetable Oil	Retailer A	6.4%	6.7%	6.9%	13.4%	16.4%
	Retailer B	8.5%	10.0%	19.4%	25.7%	26.8%

[a]2020 figures pertain to the first half of 2020 only

quarter of the previous year and the quarterly goals previously set. The draft includes various aspects of the sales promotion, such as the combinations of products and discounts that will be part of a given promotion, its duration, the target pass-through, deductions[2] and other financial costs, among others. These aspects are subject to heavy weekly negotiation. The retailer has the final say on all aspects regarding the trade promotion, with the draft being merely a suggestion. However, since it is not worthwhile for the retailer to offer a sales promotion without a solid trade promotion behind it, the retailer often negotiates with the manufacturer. The manufacturer's commercial team negotiates based on forecasts of sales and commodity prices, as well as their own expertise. After the sales promotion has ended, the manufacturer's commercial team members analyze the results using the latest sales data and their experience.

1.2 *Sales Promotion Evaluation*

To the best of the authors' knowledge, there is a gap in the literature in what concerns the evaluation of sales promotion without the use of scanner-level data, which may not be available to the manufacturer. Nevertheless, there is a large body of research on the retail side, using scanner-level and other retailer-specific data. Three works stand

[2] Retailers deduct from the manufacturer's invoice a compensation for their promotional or advertising efforts that the retailer believes to be just compensation.

out in this context. Cooper et al. [7] implemented a sales promotion forecasting system to aid a large retailer to plan promotions both efficiently and effectively. The system leverages daily or weekly scanner-level data from multiple stores and consumer panel data, as well as promotional features such as the type of ad used and the product display allocated. Forecasting was performed through linear and log-linear models. Divakar et al. [9] developed a large forecasting model inserted in a DSS for a billion-dollar revenue CPG company with extensive use of scanner-level data among other data sources. Analogously to the DSS implemented by [7], sales forecasting was performed through linear and log-linear models, using a wide array of features, including display sizes for each brand, and prices of multiple products across brands. Abolghasemi et al. [1] expanded on both of these articles, testing several models, including various time-series and machine learning models. Unlike the other previous works, less data was available. Only historical sales and price data was used. Ultimately, the authors show that the volatility of demand has a significant impact on forecasting accuracy and that simple statistical models can outperform more complex models when this volatility is present.

1.3 Models and Methods

A supervised machine learning model is one where the training data includes both the input vectors and their corresponding target vectors Bishop [2]. If there are a finite number of categories to match to each input vector, the model is performing a classification process. Otherwise, if one or more continuous variables are associated with each input vector, the model is performing a regression [2]. In a regression problem, a number of features correspond to a numeric variable, called the target, whose relationship to the features is to be approximated via a regression model. Generally, the dataset, which contains multiple data points with the aforementioned features and target, is split in three partitions, namely, the training, validation and test datasets. The training dataset is used to fit the model, iteratively, in a way that minimizes a function, called the loss function. The loss function indicates how well the model, given the corresponding features, estimates the target. However, this minimization can be accomplished to an extreme degree by a sufficiently complex model that has simply memorized the training set. This is called overfitting, meaning that the model has, in all likeliness, not learned the underlying relationship between the features and the target. An overfit model generally translates to one with poor foresight capability. To prevent this, the models are tested on a validation dataset, which assures the model has not become overfit by evaluating the model's performance on a dataset it has not been trained on. This validation is then used to select the best model among the ones trained. However, since the validated model is biased towards the validation dataset, given that it was chosen for its validation performance, the model is then evaluated on a different dataset, the test dataset. From this last evaluation, its test performance is calculated and an estimate of the model's real-world performance can be obtained [2].

Several machine learning techniques for regression exist, of which two of interest to this work are discussed below. Both techniques are based on decision trees: models which split the feature space into regions via successive recursive binary splits, each of which is governed by a simple model, usually a constant [2]. The root node begins with the whole dataset and selects the feature that best splits it, according to a given metric, dividing the space into two regions. The resulting two regions are then recursively split, until a stopping criteria has been met.

Random forests

Random forests is a learning method developed by Breiman [5] for both classification and regression problems. This method uses an ensemble of decision trees, which together form a strong model. The uniqueness of the method comes from the conjunction of random feature selection with bootstrap aggregation, also called bagging. Random feature selection restricts each split to only use a random subset of features, further reducing correlation between trees. Bagging increases the diversity of the trees trained by sampling the original training set with replacement, thereby generating different datasets.

Gradient boosting machines

A learning method developed by Friedman [10], gradient boosting machines (GBM) was initially called "multiple additive regression trees". GBM successively trains decision trees on the residuals left by the previous trees, correcting their deficiencies, which is shown by [10] to be a combination of both gradient descent and boosting. Gradient descent is an iterative algorithm for finding the local minimum of a differentiable function, by successively calculating the gradient at any point in said function and stepping in the opposite direction to it. Boosting is a meta-algorithm that combines the outputs of many weak learners into one strong learner by adding the predictions of each one of the models together to arrive at the ensemble's prediction.

2 Solution Method Proposed

The solution method proposed is a decision support system (DSS) comprised of three tools, namely, the Promotional Registry Tool (PRT), the Promotional Scenario Planner and Simulator (PSPS) and the Scenario Runner (SR). The system also encompasses the predictive models used by the SR and the database connecting the various elements together, as shown in Fig. 1. The PRT was developed to aid promotional plan data collection, making it a less error-prone process and streamlining future data flows, assuring the system stays up-to-date. The PSPS is the focal interface of the DSS, with the main purpose of displaying the results of the simulation of

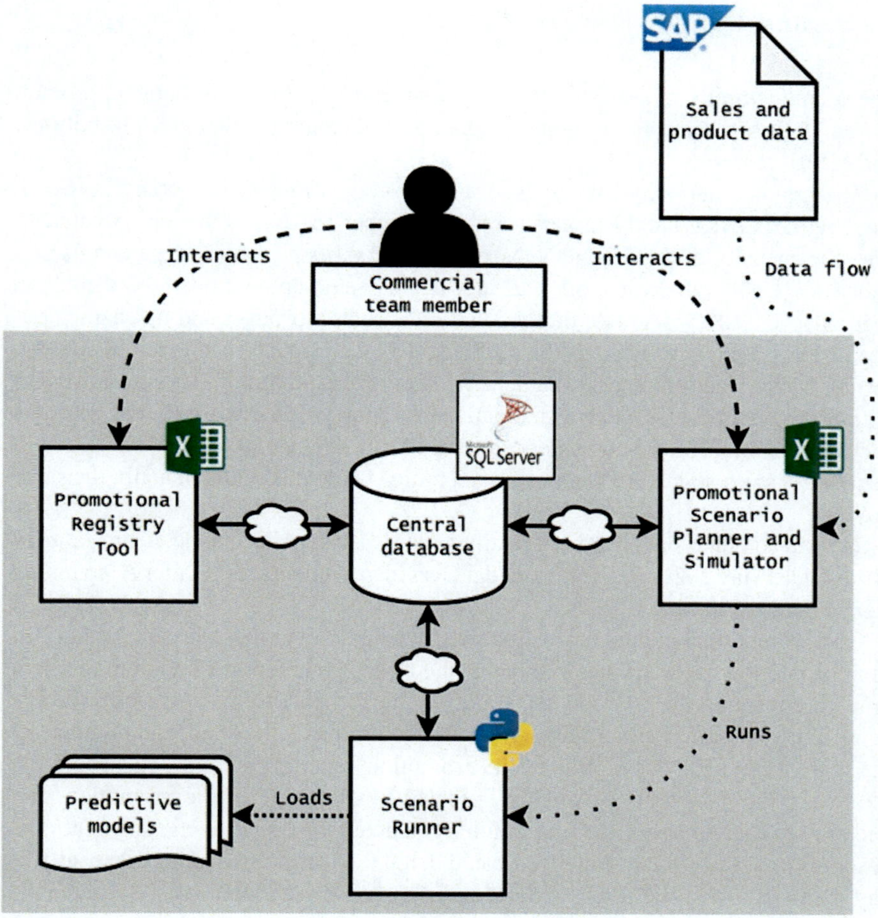

Fig. 1 Decision support system diagram

future promotional scenarios, powered by the SR. The latter generates these results after loading and running the previously trained gradient boosting machine models, which are the focus of this work. These models need to balance both predictive performance and sensitivity to some key features, since the user is likely to often want to experiment with the system in atypical ways, for example, with historically rare discounts.

2.1 Data

The manufacturer has available four key sources of information, namely, product hierarchy data, sales data, promotional plan data, including competitor promotions, and MSRP tables.

The product hierarchy data details for each product involved its brand, its Stock Keeping Unit (SKU) and European Article Number (EAN), as well as its capacity (i.e., the capacity of the product's package in liters), its packaging type and its category (vegetable oil or olive oil) and subcategory (e.g., organic olive oil, sunflower seed oil). The SKU is a code that uniquely identifies a product and its characteristics inside a company, whereas the EAN identifies a product in an external context, serving as the code for use by retailers or other external entities. The same product with different attributes, such as thematic packaging, will have a different SKU for each combination, so the company can track its different variations. However, it will generally have a single EAN, so external entities know it is fundamentally the same product.

In terms of sales data, for both retailers, daily sell-in data spanned from January 2011 to late May 2021. Daily sell-out data was also available for retailer A, spanning from July 2016 to March 2021.

The promotional plan data was recorded manually over the years, including, for each promotion, the start and end dates, the type of marketing campaign, the geographic coverage, the promotional price (if the campaign targets a single EAN), as well as two human descriptions, one for the products involved[3] and another for the promotion.[4] The description of the promotion generally started with specifying the marketing campaign, optionally followed by the name of the promotion, and ended with a summary of the promotion's characteristics, such as the discount. The description of the products involved could detail the brand, subcategory, capacity or any combination of the three. When analyzing company promotions, these details enabled the creation of a list of potential SKUs, using the product hierarchy data. This was necessary due to the lack of an explicit listing of SKUs included in a given promotion. The list of potential SKUs was then filtered, removing inactive SKUs (that is, SKUs which had zero total sales during the 6 months prior to the promotion in hand). Finally, by using the EAN associated with each SKU, we arrived at a list of EANs believed to have been included in the promotion.

2.2 Mathematical Notation and Formulas

For clarity, this section introduces the main notation (Table 3) and formulas (Eqs. 1 to 15) applied throughout this paper.

[3] For example: "Extra Virgin Olive Oil".

[4] For example: "Pamphlet: "fill up your pantry": 35% discount on a selection of gourmet Brand A olive oil".

Table 3 Main notation

Parameter	Description
p	Promotion
SD_p	Start date of a promotion
ED_p	End date of a promotion
$\mathrm{STK}_{\mathrm{start}}$	No. of days between a promotion's start date and the start of its stocking period
$\mathrm{STK}_{\mathrm{end}}$	No. of days between a promotion's start date and the end of its stocking period
e	EAN
$\mathscr{E}_p$	Set of EANs included in promotion p
$\mathscr{C}_p$	Set of EANs included in promotion p's product category
$D_{e,p}$	Discount offered on EAN e during promotion p
$L_{e,d}$	Sales of EAN e on day d in liters

Total liters sold during promotion p:

$$L_p = \sum_{d \in \mathscr{D}_p} \sum_{e \in \mathscr{E}_p} L_{e,d} \tag{1}$$

where the days considered in promotion p are given by:

$$\mathscr{D}_p = \{\mathrm{SD}_p, \cdots, \mathrm{ED}_p\} \quad \text{(Sell-out version)} \tag{2}$$

$$\mathscr{D}_p = \{\mathrm{SD}_p - \mathrm{STK}_{\mathrm{start}}, \cdots, \mathrm{SD}_p - \mathrm{STK}_{\mathrm{end}}\} \quad \text{(Sell-in version)} \tag{3}$$

Average daily sales of EAN e on day d in liters:

$$L_{\mathrm{avg}_{e,p}} = \frac{\sum_{d' \in \mathscr{D}_d} L_{e,d'}}{d_{\max} - d_{\min} + 1}, \quad \mathscr{D}_d = \{d - 13 \text{ months}, \cdots, d - 1 \text{ month}\} \tag{4}$$

where

$$d_{\min} = \underset{d}{\operatorname{argmin}}\, L_{e,d}, \quad d \in \mathscr{D}_d, \quad L_{e,d} > 0 \tag{5}$$

$$d_{\max} = \underset{d}{\operatorname{argmax}}\, L_{e,d}, \quad d \in \mathscr{D}_d, \quad L_{e,d} > 0 \tag{6}$$

Promotional multiplier of promotion p:

$$\mathrm{PM}_p = \frac{L_p}{(\mathrm{ED}_p - \mathrm{SD}_p + 1) \cdot \sum_{e \in \mathscr{E}_p} L_{\mathrm{avg}_{e,\mathrm{SD}_p}}} \tag{7}$$

Estimating the total liters sold during promotion p:

$$\hat{L}_p = (\mathrm{ED}_p - \mathrm{SD}_p + 1) \cdot \sum_{e \in \mathscr{E}_p} \hat{L}_{\mathrm{avg}_{e,p}} \tag{8}$$

Estimating the promotional multiplier of promotion p:

$$\hat{\mathrm{PM}}_p = \frac{\sum_{e \in \mathscr{E}_p} \hat{L}_{\mathrm{avg}_{e,p}}}{\sum_{e \in \mathscr{E}_p} L_{\mathrm{avg}_{e,\mathrm{SD}_p}}} \tag{9}$$

Weight of EAN e on promotion p:

$$\mathrm{WEP}_{e,p} = \frac{L_{\mathrm{avg}_{e,p}}}{\sum_{e' \in \mathscr{E}_p} L_{\mathrm{avg}_{e',p}}} \tag{10}$$

Weight of promotion p on its category:

$$\mathrm{WPC}_p = \frac{\sum_{e \in \mathscr{E}_p} L_{\mathrm{avg}_{e,p}}}{\sum_{e \in \mathscr{C}_p} L_{\mathrm{avg}_{e,p}}} \tag{11}$$

Set of promotions active during day d:

$$\mathscr{P}_d = \{p' : \mathrm{SD}_{p'} \leq d \leq \mathrm{ED}_{p'}, \quad \forall p'\} \tag{12}$$

Cannibalization effects on EAN e during promotion p:

$$\mathrm{CN}_{e,p} = \sum_{d \in \{\mathrm{SD}_p, \cdots, \mathrm{ED}_p\}} \sum_{e' \in \mathscr{C}_p \setminus \{e\}} \max_{p' \in \mathscr{P}_d} (D_{e',p'} \cdot \mathrm{WEP}_{e',p'} \cdot \mathrm{WPC}_{p'}) \tag{13}$$

Promotional intensity of EAN e during promotion p:

$$\mathrm{PI}_{e,p} = \sum_{d \in \{\mathrm{SD}_p - 30, \cdots, \mathrm{SD}_p\}} \max_{p' \in \mathscr{P}_d} (D_{e,p'} \cdot \mathrm{WEP}_{e,p'} \cdot \mathrm{WPC}_{p'}) \tag{14}$$

Promotional intensity of the category during promotion p:

$$\mathrm{PIC}_p = \sum_{d \in \{\mathrm{SD}_p - 30, \cdots, \mathrm{SD}_p\}} \sum_{e \in \mathscr{C}_p} \max_{p' \in \mathscr{P}_d} (D_{e,p'} \cdot \mathrm{WEP}_{e,p'} \cdot \mathrm{WPC}_{p'}) \tag{15}$$

2.2.1 A Note on Determining the Sales of a Promotion

The method for calculating L_p using sell-out data is intuitive, since in this case $L_{e,d}$ represents the sales of EAN e on day d to customers (see Eq. 1 and Eq. 2). In the sell-in case, given by Eq. 3, the total sales are approximated by the liters that arrived to the retailer during the stocking period, with $L_{e,d}$ representing the liters of EAN e delivered to the retailer on day d, from which customers will buy at an uncertain date. This stocking period is defined by $\text{STK}_{\text{start}}$ and STK_{end}, which are normally 7 and 1, respectively, to encompass the week before the promotion. This approximation holds well for higher rotation products that have a fairly predictable and strong demand, since the retailer is incentivized to purchase only the necessary amount it is anticipating to sell, as stocking more than a few weeks' worth of these items is undesirable and costly. However, this approximation does not hold particularly well for low-rotation items, for which the retailer is incentivized to stock only a small initial amount, often at a reduced price, that sells over a long period of time.

Turning to expert advice, following talks with the manufacturer's commercial team, the stocking period was set as the week before the promotion. If the retailer stockpiles heavily during the stocking period or before, peaks and troughs of perceived promotional sales occur, respectively, translating into extreme promotional multipliers, that is, selling below half or above ten times the normally expected amount. This was corrected by filtering out promotions whose promotional multipliers were outside these limits, which were set by the commercial team.

2.3 *Feature Engineering*

To better equip the developed GBM regression models, which are discussed in Sect. 2.6, a number of features were created on top of the basic features available. However, some features showed more significance than others, as is touched upon in Sect. 3.

The features *Weight of EAN e on promotion p* and *Weight of promotion p on the category* capture the effects related to the weight in sales of an EAN on the promotion (Eq. 10) and the weight of a promotion on its category (Eq. 11), respectively. This allows distinguishing, for instance, best-sellers from niche products, and high impact promotions from low impact promotions, respectively.

Cannibalization attempts to capture the effect that other concurrently promoted manufacturer-owned EANs have on a given EAN e during promotion p (Eq. 13).

The *Promotional intensity of an EAN* (Eq. 14) and the *Promotional intensity of the category* (Eq. 15) features enable the model to take into consideration past promotions of a given EAN or category, respectively, via a weighted sum of past discounts, for each day in the month previous to the current promotion.

Since discount and promotional policies are dependent on the current price of the commodity underlying the category, as well as forecasts of this commodity, two external features were included, namely, the *Commodity monthly price*, in United

States dollars per metric ton, and the *Commodity monthly percentage price change*. The commodities chosen were the global price of olive oil and the global price of sunflower oil, for the olive oil and the vegetable oil categories, respectively, and the data in question was sourced from the Federal Reserve Bank of St. Louis.[5]

Other features were also created, but preliminary tests showed minimal impacts on the models. Therefore, they are not described in detail here.

2.4 Model Target Definition

Defining the target for the model was not straightforward, as there were many possibilities to explore, given the peculiarity of the high promotional frequency present in both categories and for both retailers. This intensity hinders a traditional analysis, which would involve calculating a sales lift on top of a baseline estimate. In fact, with more promotions, less non-promoted data points are available for deriving such an estimate [3]. The proposed method abandons the concept of sales lift and instead focuses on calculating the performance of a given promotion as a multiple of its average sales, called the promotional multiplier (PM), as presented in Sect. 2.2.

With the goal of estimating L_p, the models' target is then $L_{\text{avg}_{e,p}}$, and therefore generating EAN-level predictions. The models operate at the EAN-level in order to be able to capture individual EAN effects. To ultimately arrive at promotion-level predictions, the various EAN-level predictions are aggregated and transformed into an estimate of the promotional multiplier of promotion p, $\hat{\text{PM}}_p$, as defined in Eq. 8. Additionally, estimating $L_{\text{avg}_{e,p}}$ rather than $L_{e,p}$ reduces the complexity of the model by removing the issue of dealing with how the length of a promotion affects its total sales.

2.5 Metrics for Model Evaluation

Several accuracy methods (or metrics) exist to evaluate and compare different models, of which three standard and widely used ones were chosen, namely, the mean absolute percentage error (MAPE), the mean percentage error (MPE), also called bias, and the R-squared, R^2, for both EAN- and promotion-level predictions. The latter metric, also called the coefficient of determination, assesses the goodness of fit of a given model. This assessment is performed by comparing the squared residuals of the model's predictions for a given dataset with the squared residuals of produced by a model that predicts the average of said dataset [8]. In fact, a model that predicts the average of a dataset would have a R^2 value of 0, whereas a worse model than this would have a negative R^2 value and an "ideal" model would have a R^2 value of 1 [6]. Additionally, to differentiate between errors on low-selling items and high-selling

[5] https://fred.stlouisfed.org.

items, weighted metrics are used, namely, WMAPE and WMPE. Equations for each of these five metrics are shown below, where the prediction $\hat{y}_n$ is compared to the actual value y_n, and $\bar{y}$ corresponds to the average value to be predicted.

$$\text{MAPE}\ (\%) = \frac{100}{N} \sum_n \left| \frac{\hat{y}_n - y_n}{y_n} \right| \tag{16}$$

$$\text{MPE}\ (\%) = \frac{100}{N} \sum_n \frac{\hat{y}_n - y_n}{y_n} \tag{17}$$

$$R^2 = 1 - \frac{\sum_n (y_n - \hat{y}_n)^2}{\sum_n (y_n - \bar{y})^2} \tag{18}$$

$$\text{WMAPE}\ (\%) = 100 \cdot \frac{\sum_n \left| \frac{\hat{y}_n - y_n}{y_n} \right| \cdot W_n}{\sum_n W_n} \tag{19}$$

$$\text{WMPE}\ (\%) = 100 \cdot \frac{\sum_n \frac{\hat{y}_n - y_n}{y_n} \cdot W_n}{\sum_n W_n} \tag{20}$$

The weight used in WMAPE and WMPE, W_n, is the *Average daily sales* of each EAN when evaluating EAN-level predictions, whereas for promotion-level predictions it is the total *Average daily sales* of the EANs involved in the promotion.

2.6 Models

Two types of model were created, namely, the α-type and the β-type models, one for each retailer-category pair. Both types of model used the Gradient Boosting Machine (GBM) learning method, with hyperparameter tuning of the maximum depth of each individual tree and the number of trees. In preliminary tests, GBM and some other machine learning algorithms were experimented with, where the former was selected due to its overall best performance in this context. The latter included linear regressions, random forests and deep learning models, among others, all of which are widely used both in the academic literature and in practice, as well as readily available. Similarly, a standard loss function, was used, as described in Eq. 21, where the model's estimate, $y(\vec{x})$, is compared to the target value t.

$$L(t, y(\vec{x})) = C(y(\vec{x}) - t)^2, C > 0 \tag{21}$$

α-type models were generated via a modification of a feature selection algorithm introduced by [12]. This algorithm automatically selects features while accounting for the bias introduced by ranking the feature set on the training set. Hyperparameter tuning was included in the same bias avoiding spirit, as detailed in Algorithm 1. After analyzing the behavior of model α-type models while embedded in the PSPS and its scenario evaluation results, their extreme lack of sensitivity became clear.

That is, these models did not change their predictions according to user changes in critical variables, such as discount percentage. The sensitivity of a model is a critical aspect to assure the quality of the feedback given to the user during the planning and simulation process.

To overcome this issue presented by α-type models, β-type models were developed. In this iterative process, there was a particular focus on the resulting scatter plot of the predicted versus the actual values of the promotional multiplier for the several data sets. Only the GBM algorithm and dataset splits were kept the same, with the automatic feature selection algorithm having been put aside. For both α- and β-type models, the training set spans from the start of 2015 until the end of 2018, with the validation set encompassing the whole of 2019, and the testing set spanning 2020. The β-type models' hyperparameters came from expert advice, with its maximum depth and number of trees fixed at 5 and 50, respectively. The features were selected iteratively and based on expert knowledge, starting with a few nonnegotiable ones (namely, the discount percentage and the month of the promotion at hand), as well as the segment, promotional price, capacity and brand of the EAN in question. The latter was only valid for the vegetable oil category since the manufacturer has more than one prominent name brand for the market in question. From this base group of features, the remaining were added and discussed, comparing their effect on training and validation metrics, and a few were ultimately selected. The feature *Average daily sales* was removed from consideration, since it negatively affected the model's sensitivity, overshadowing other key features in importance upon its addition. Additionally, the external variables were discarded, given how minute the benefit the models were able to extract from their usage was, as can be later seen in Tables 4 and 5. Moreover, adding these variables to the final model would make the model more complex, as well as a more complex data flow, given that these variables would need to be updated on a monthly basis.

When the SR runs the models, the resulting estimated promotional multipliers are capped at a minimum of 0.5 and a maximum of 10, in order to avoid possible outlier effects, as mentioned in Sect. 2.2.1.

3 Results

In this section, the different models' results and metrics are analyzed and compared, from which relevant conclusions are drawn about the product categories and retailers. The system deployment is discussed, as well as how the data available impacted the work.

Algorithm 1 Model tuning algorithm.

Divide dataset into training (trs), validation (vls) and testing datasets
for i = 1 to n models **do**
    $trs'_n \leftarrow$ bootstrap resample of trs
    $vls'_n \leftarrow$ bootstrap resample of vls
    Train temporary Random Forest model (RF) on trs'_n, with *ntrees*=500, *max_depth*=3
    $FI_n \leftarrow$ RF feature importance
    for *hc* **in** hyperparameter combinations **do**
        FL (feature list) $\leftarrow FI_n$
        while FL $\neq \emptyset$ **do**
            Fit a new GBM model on trs'_n using FL and *hc*
            Apply model to predict vls'_n
            Save model and its results for future reference
            FL $\leftarrow$ FL \{least important feature according to FI_n}
Select model with best vls′ prediction performance, use its feature subset and *hc* to train a final GBM on both trs and vls
Use final model to predict the testing set, where its performance is the real world performance estimate

3.1 *Feature Selection*

The feature selection results for α-type models are shown in Table 4, where the multiple feature importances, expressed as percentages, are ranked. Unranked features are features which were not selected for use for a given model. Here it is clear that feature *Average daily sales* dominates every model, except for the olive oil model for retailer A, where the *Weight of the promotion on the category* surpasses it. This is not surprising since feature *Average daily sales* naturally serves as an adequate starting point for estimating how much a given EAN will sell on a daily basis in the next promotion. However, in practice it appears to have a limiting effect on the models' output range, as shown in the scatter plots of Fig. 2. Many of the company promotion features are overshadowed by others that encapsulate the promotional context more meaningfully and are more easily comprehended by the models. It is apparent how α-type models overall give no weight to key features, namely, the month and the segment. The *Discount*, *Promotional price* and *Promotional price per liter* features are dependent on one another, such that α-type models naturally split the importance among them. The feature importances of α-type models also show us that not a single competitor promotion feature was highlighted by the model in a significant way, possibly pointing to a serious shortcoming in capturing competition effects.

Alongside the feature selection, an exhaustive grid search was performed for each model fitted, on the two hyperparameters mentioned in Sect. 2.6, namely, for a *max_depth* between 3 and 6, and for *ntrees* of 50, 60, 70, 80, 90 or 100. Ultimately, all α-type models used a maximum depth of 3, with the number of trees ranging from 70 to 100.

The features selected for the β-type models are shown in Table 5, having been selected according to the method detailed in section 2.6.

Table 4 Feature importances for the various α-type models

Feature importance	Olive oil				Vegetable Oil			
	Retailer A		Retailer B		Retailer A		Retailer B	
	%	Rank	%	Rank	%	Rank	%	Rank
Average daily sales	37.28	2	76.93	1	77.95	1	84.34	1
Brand	–	–	–	–	–	–	–	–
Cannibalization	–	–	2.91	5	1.98	6	–	–
Capacity	–	–	0.04	16	–	–	–	–
Commodity monthly percentage price change	–	–	0.41	15	–	–	–	–
Commodity monthly price	–	–	1.73	7	–	–	–	–
Competitor promotional intensity	–	–	–	–	–	–	–	–
Days elapsed since last competitor promotion	–	–	–	–	–	–	–	–
Days elapsed since last promotion	–	–	–	–	–	–	–	–
Discount	–	–	0.76	9	–	–	–	–
Duration	–	–	0.44	13	–	–	–	–
MSRP	–	–	0.74	10	–	–	0.93	7
Month	–	–	0.03	17	–	–	–	–
Number of concurrent promotions	–	–	0.62	11	–	–	–	–
Number of recent competitor promotions	–	–	–	–	–	–	–	–
Number of recent promotions	–	–	–	–	–	–	–	–
Percentage of EANs included in the promotion	5.17	3	0.00	18	–	–	0.36	9
Promotional intensity of an EAN	3.88	5	3.70	3	3.29	4	2.18	5
Promotional intensity of the category	–	–	1.25	8	–	–	3.66	3
Promotional price	–	–	1.89	6	–	–	1.44	6
Promotional price per liter	–	–	4.54	2	4.89	3	0.78	8
Recent competitor promotional intensity	–	–	–	–	–	–	–	–
Segment	–	–	0.43	14	–	–	–	–
Weight of an EAN on the promotion	49.05	1	2.97	4	9.14	2	3.96	2
Weight of the promotion on the category	4.62	4	0.61	12	2.75	5	2.35	4

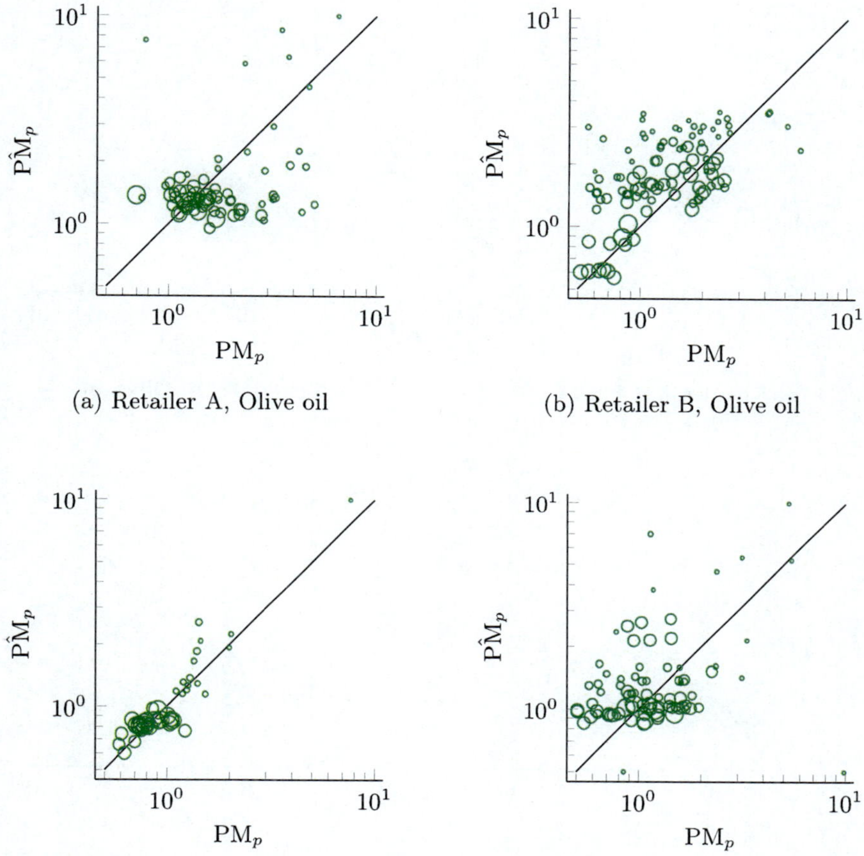

(a) Retailer A, Olive oil

(b) Retailer B, Olive oil

(c) Retailer A, Vegetable oil

(d) Retailer B, Vegetable oil

Fig. 2 Promotion-level α-type model scatter plots for both categories and retailers, showing the relationship between actual versus predicted values of promotional multiplier

3.2 Comparing Model Results

Comparing the metrics for both types of models, β-type models come out as superior to their α-type alternatives, in all but a few metrics in some specific models, as shown in Table 6. Overall, β-type models reveal better performance on the testing dataset, possess stronger R^2 values and, these both being correlated, less disperse scatter plots, when compared to their counterpart, as shown in Fig. 3. The α-type models tend to mostly predict values around 1. This means that the model predicts the overwhelming majority of promotions to be close to average, which translates into poor feedback for the planning process. β-type models are in practice more responsive and fit the whole response range better. Although α-type models are generally not far behind their counterparts when it comes to promotion-level metrics, their low performance

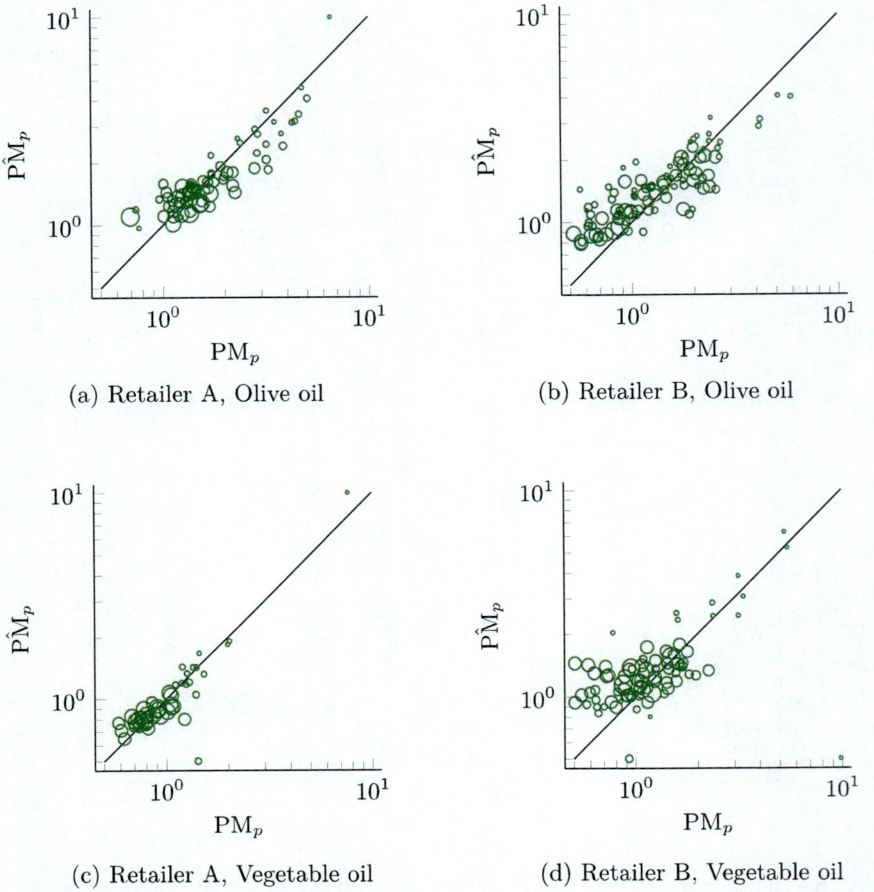

Fig. 3 Promotion-level β-type model scatter plots for both categories and retailers, showing the relationship between actual versus predicted values of promotional multiplier

is apparent at the EAN-level, as evidenced by their EAN-level MAPE metrics. This is due to the models' inadequate grasp of low-selling products, indicated by comparison with the more redeeming WMAPE metric, which places a heavier focus on high-selling products.

3.3 Comparing Categories

Of both categories, the vegetable oil category should theoretically be easier to predict given the weaker competition. This is suggested by the leading market share held by the company, which translates into less relevant competition effects that are hard to

Table 5 Feature importances for the various β-type models

Feature importance	Olive oil				Vegetable Oil			
	Retailer A		Retailer B		Retailer A		Retailer B	
	%	Rank	%	Rank	%	Rank	%	Rank
Average daily sales	–	–	–	–	–	–	–	–
Brand	–	–	–	–	0.05	9	0.02	9
Cannibalization	–	–	–	–	18.69	2	5.81	4
Capacity	1.76	6	12.47	2	1.88	6	1.91	6
Commodity monthly percentage price change	–	–	–	–	–	–	–	–
Commodity monthly price	–	–	–	–	–	–	–	–
Competitor promotional intensity	–	–	–	–	–	–	–	–
Days elapsed since last competitor promotion	–	–	–	–	–	–	–	–
Days elapsed since last promotion	–	–	–	–	–	–	–	–
Discount	1.19	7	1.57	7	0.65	7	0.49	8
Duration	–	–	–	–	–	–	–	–
MSRP	–	–	–	–	–	–	–	–
Month	1.05	8	0.92	8	0.42	8	1.47	7
Number of concurrent promotions	–	–	–	–	–	–	–	–
Number of recent competitor promotions	–	–	–	–	–	–	–	–
Number of recent promotions	–	–	–	–	–	–	–	–
Percentage of EANs included in the promotion	–	–	–	–	–	–	–	–
Promotional intensity of an EAN	74.42	1	65.57	1	71.38	1	68.84	1
Promotional intensity of the category	2.32	5	3.62	5	–	–	–	–
Promotional price	3.91	4	5.16	4	1.92	5	3.04	5
Promotional price per liter	–	–	–	–	–	–	–	–
Recent competitor promotional intensity	–	–	–	–	–	–	–	–
Segment	5.81	3	7.47	3	2.15	4	8.37	3
Weight of an EAN on the promotion	–	–	–	–	–	–	–	–
Weight of the promotion on the category	9.54	2	3.23	6	2.86	3	10.05	2

Table 6 Metrics for the various α- and β-type models

	Metrics	Olive oil				Vegetable oil			
		Retailer A		Retailer B		Retailer A		Retailer B	
		α	β	α	β	α	β	α	β
EAN-level	MAPE	208.38	55.68	91.57	48.98	301.91	40.78	310.59	52.50
	MPE	187.15	24.61	70.72	29.84	289.06	31.20	281.88	33.33
	R^2	0.6956	0.9000	0.6805	0.7880	0.9207	0.9511	0.7666	0.6818
	WMAPE	34.69	21.45	53.99	37.52	15.28	11.16	44.72	42.68
	WMPE	0.68	3.01	36.52	22.75	1.59	0.67	22.51	30.40
Promotion-level	MAPE	43.80	16.92	62.78	29.22	12.77	10.37	54.84	34.49
	MPE	6.87	−0.70	51.38	16.84	3.30	−0.45	31.12	21.04
	R^2	−0.7778	0.7015	−0.0862	0.7240	0.8290	0.8527	−0.4358	0.2282
	WMAPE	26.34	16.09	38.65	25.68	10.88	9.04	41.34	35.46
	WMPE	−4.48	−0.78	25.30	14.95	0.16	−0.03	19.99	26.55

capture presently with the data available. This is a possible reason why the *Cannibalization* feature takes on significance in retailer A's vegetable oil β-type model. In the presence of weak competitor products, the fiercest competition is then to be found in the company's other products. For retailer B, this is not immediately apparent as both model types struggled to capture *Cannibalization*'s significance, possibly due to the sell-in noise, non-apparent differences between the retailers, or a combination of factors.

The olive oil category, more heavily contested, is naturally more dependent on the actions of the competition. However, no feature adequately encapsulated this effect, as noticed experimentally, during the selection of features for β-type models, in which no competitor effect related features added value in a significant way to any model.

3.4 Comparing Retailers

Retailer B is harder to predict than retailer A, despite the former having a larger dataset. This could be attributed to differences between the retailers, specifically differences in the products offered for each category. Fig. 4 shows the Pareto[6] distribution for the company's products in terms of liters sold, demonstrating the wider variety of products offered by retailer B in each category, one of the traits the retailer is known for.

However, a stronger justification is the fact that retailer B's models are trained on a sell-in approximation of the sell-out numbers, which introduces noise in the data that

[6] Vilfredo Pareto, a famous Italian polymath, coined the Pareto principle after observing that 80% of Italian land was owned by 20% of its population. The principle essentially states that a minority of agents bring about a majority of consequences.

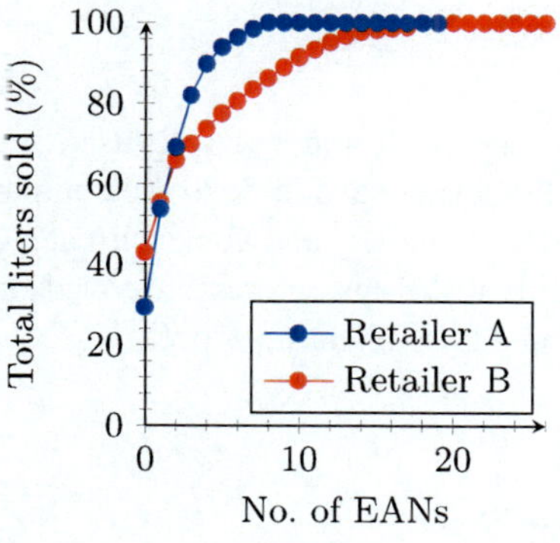

(a) Olive oil

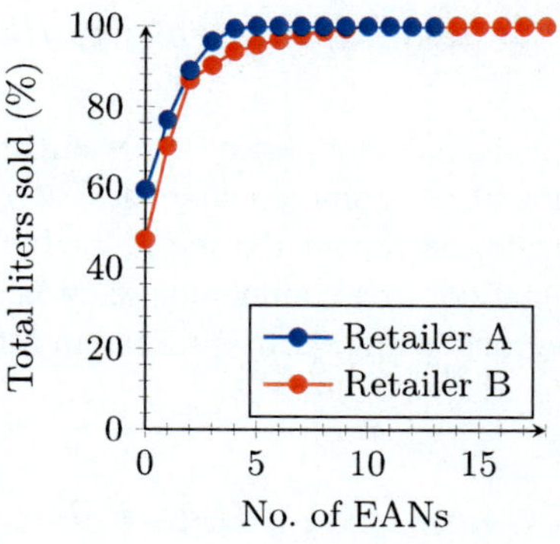

(b) Vegetable oil

Fig. 4 Pareto distributions of liters sold by the company, for both retailers and categories

Table 7 β-type model metrics for retailer A and retailer A*

	Metrics	Olive oil		Vegetable oil	
		Retailer A	Retailer A*	Retailer A	Retailer A*
EAN-level	MAPE	55.68	72.40	40.78	106.50
	MPE	24.61	4.17	31.20	24.28
	R^2	0.9000	0.4694	0.9511	−0.1010
	WMAPE	21.45	51.59	11.16	77.61
	WMPE	3.01	−23.94	0.67	−62.15
Promotion-level	MAPE	16.92	39.68	10.37	54.63
	MPE	−0.70	−19.12	−0.45	−17.24
	R^2	0.7015	−0.2183	0.8527	0.1739
	WMAPE	16.09	32.20	9.04	44.00
	WMPE	−0.78	−22.86	−0.03	−39.84

naturally degrades the models' performance. To validate such a hypothesis, given that sell-in data is naturally available to all retailers, the sell-in version of retailer A, referred to as retailer A*, was used to train β-type models in order to analyze the effects of the sell-in approximation. The metrics of the β-type retailer A* models are shown in Table 7, supporting the hypothesis that a large portion of the difference between retailer A and retailer B's model performance is explained by the difference between the quality of their sales data.

3.5 System Deployment Results

The full decision support system was deployed successfully and is actively used by the manufacturer's commercial team. The β-type models are available to the company, alongside the rest of the tools, for both retailer A and B, and for each category combination. The commercial team has stated that the system has improved their promotional plan registry process and streamlined their planning process.

3.6 Comments on Data-Related Impacts

The sales data available to the manufacturer pales in comparison to the wealth of information the large retailer possesses, which involve scanner-level data, often going household-level deep, in the case of retailers with extensive and successful loyalty programs. Such data, besides allowing its holder to understand which products were sold when and at what price point, enables also the tracking of consumer purchase patterns. Furthermore, it helps remove some uncertainty over the promotional plan's execution and reach, since sales could be directly linked to a particular promotion, especially for promotions of restricted geographic coverage. This data could be of great value to the manufacturer, in order to generate better forecasts and better negotiate future trade promotions, as well as possibly helping to predict stockpiling effects and timing post-stocking demand.

4 Conclusions

This work aimed to evaluate trade promotions from the point of view of a consumer packaged goods manufacturer, with limited and unpolished data. This enables the manufacturer's commercial team to avoid below-average promotions and empowers them to better negotiate trade promotions with the retailers, especially given the high frequency with which promotions are currently being held. To this effect, a comprehensive and tailored decision support system was developed, to allow the team to plan and simulate specific promotional plans, receiving an estimate of the resulting sales, which then can be used as intelligence for negotiation with the retailers. This work stresses the importance of having access to quality data, which most consumer packaged goods companies do not have. By acting as middlemen to the manufacturers, retailers are granted access to a wealth of data pertaining to the customer base of the manufacturer, as well as their purchase patterns, which could be of great use to the manufacturer. Of the retailers that deal with the manufacturer, the one that provided sell-out data fueled the best models obtained, testifying to the importance of such data's availability. Both parties could come together and share information to improve their competitiveness and strengthen their partnership. A tighter manufacturer-retailer relationship could mean more efficient management of inventory and stronger margins for both.

The literature covering the forecasting of sales induced by trade promotions from the manufacturer's perspective is rather scarce when compared to the literature regarding the concerns of the retailer. This comes naturally as a consequence of the retailers' significant push for research and development. As retailers often face fierce competition, they seek to leverage the big amount of data they possess into competitive advantage, funding endeavors in the field of sales forecasting and inventory management. In this context, this work contributes to the literature and helps to ease the gap in it, by exploring the themes discussed from the manufacturer's point-of-view.

An interesting avenue not explored by this work lies in more effectively categorizing and distinguishing products to improve EAN-level predictions and allow the methodology to be extended, covering manufacturers wider and more diverse product portfolios in a given category. This work explored predicting sales for a given product-promotion pair, such that exploring different targets, at diverse granularities, and their trade-offs could prove to be another interesting future research avenue.

References

1. Abolghasemi, M., Beh, E., Tarr, G., Gerlach, R.: Demand forecasting in supply chain: The impact of demand volatility in the presence of promotion. Comput. Ind. Eng. **142**, 106380 (2020)
2. Bishop, C.M.: Pattern Recognition and Machine Learning (Information Science and Statistics). Springer, Berlin, Heidelberg (2006)
3. Blattberg, R.C., Kim, B., Ye, J.: Defining baseline sales in a competitive environment. Seoul J. Bus. **2**
4. Blattberg, R.C., Neslin, S.A.: Sales promotion: the long and the short of it. Market. Lett. **1**(1), 81–97 (1989)
5. Breiman, L.: Random forests. Mach. Learn. **45**(1), 5–32 (2001)
6. Chicco, D., Warrens, M.J., Jurman, G.: The coefficient of determination R-squared is more informative than SMAPE, MAE, MAPE, MSE and RMSE in regression analysis evaluation. Peer J. Comput. Sci. **7**, e623 (2021)
7. Cooper, L.G., Baron, P., Levy, W., Swisher, M., Gogos, P.: PromoCast^{TM}: a new forecasting method for promotion planning. Market. Sci. **18**(3), 301–316 (1999)
8. Devore, J.: Probability and Statistics for Engineering and the Sciences. Brooks/Cole Publishing Company (1987)
9. Divakar, S., Ratchford, B.T., Shankar, V.: Practice Prize Article- CHAN4CAST: a multichannel, multiregion sales forecasting model and decision support system for consumer packaged goods. Market Sci. **24**(3), 334–350 (2005)
10. Friedman, J.H.: Greedy function approximation: a gradient boosting machine. Ann. Stat. **29**(5), 1189–1232 (2001). Publisher: Institute of Mathematical Statistics
11. Hill, R.: Promotions: Do You Know What You Don't Know?
12. Kuhn, M., Johnson, K.: Feature Engineering and Selection: A Practical Approach for Predictive Models. CRC Press (2019)
13. Nielsen: 3,5 mil milhões de euros na "Selva Promocional" dos bens de grande consumo (2020). https://web.archive.org/web/20200920123528/. https://www.nielsen.com/pt/pt/insights/article/2019/3-point-5-billion-euros-in-promotional-jungle-of-consumer-goods/. Accessed 20 June 2021